The 1984
Olympic Scientific
Congress
Proceedings
Volume 4

Sport and Human Genetics

Series Editors:

Jan Broekhoff, PhD
Michael J. Ellis, PhD
Dan G. Tripps, PhD

*University of Oregon
Eugene, Oregon*

The 1984
Olympic Scientific
Congress
Proceedings
Volume 4

Sport and Human Genetics

Robert M. Malina and Claude Bouchard
Editors

Human Kinetics Publishers, Inc.
Champaign, Illinois

Library of Congress Cataloging-in-Publication Data

Olympic Scientific Congress (1984 : Eugene, Or.)
    Sport and human genetics.

    (1984 Olympic Scientific Congress proceedings ; v. 4)
    Bibliography: p.
    1. Sports—Physiological aspects—Congresses.
2. Human genetics—Congresses.   3. Somatotypes—
Congresses.   I. Malina, Robert M.   II. Bouchard, Claude.
III. Title.   IV. Series: Olympic Scientific Congress
(1984 : Eugene, Or.).   1984 Olympic Scientific Congress
proceedings ; v. 4.
GV565.O46   1984 vol. 4   796 s   85-18116
[RC1235]   [612′.044]
ISBN 0-87322-011-0   68153

Managing Editor: Susan Wilmoth, PhD
Developmental Editor: Gwen Steigelman, PhD
Production Director: Sara Chilton
Copyeditor: Peg Goyette
Typesetters: Aurora Garcia and Sandra Meier
Text Layout: Cyndi Barnes
Cover Design and Layout: Jack Davis
Printed By: Braun-Brumfield, Inc.

ISBN: 0-87322-006-4   (10 Volume Set)
ISBN: 0-87322-011-0

Printed in the United States of America

10 9 8 7 6 5 4 3 2 1

Human Kinetics Publishers, Inc.
Box 5076, Champaign, IL 61820

Contents

siblings?

Series Acknowledgments

The Congress organizers realize that an event as large and complex as the 1984 Olympic Scientific Congress could not have come to fruition without the help of literally hundreds of organizations and individuals. Under the patronage of UNESCO, the Congress united in sponsorship and cooperation no fewer than 64 national and international associations and organizations. Some 50 representatives of associations helped with the organization of the scientific and associative programs by coordinating individual sessions. The cities of Eugene and Springfield yielded more than 400 volunteers who donated their time to make certain that the multitude of Congress functions would progress without major mishaps. To all these organizations and individuals, the organizers express their gratitude.

A special word of thanks must also be directed to the major sponsors of the Congress: the International Council of Sport Science and Physical Education (ICSSPE), the United States Olympic Committee (USOC), the International Council on Health, Physical Education and Recreation (ICHPER), and the American Alliance for Health, Physical Education, Recreation and Dance (AAHPERD). Last but not least, the organizers wish to acknowledge the invaluable assistance of the International Olympic Committee (IOC) and its president, Honorable Juan Antonio Samaranch. President Samaranch made Congress history by his official opening address in Eugene on July 19, 1984. The IOC further helped the Congress with a generous donation toward the publication of the Congress papers. Without this donation it would have been impossible to make the proceedings available in this form.

Finally, the series editors wish to express their thanks to the volume editors who selected and edited the papers from each program of the Congress. Special thanks go to Robert M. Malina of the University of Texas at Austin and Claude Bouchard of the Cité Universitaire of Quebec for their work on this volume.

Jan Broekhoff,
Michael J. Ellis, and
Dan G. Tripps

Series Editors

Series Preface

Sport and Human Genetics contains selected proceedings from this inter-disciplinary program of the 1984 Olympic Scientific Congress, which was held at the University of Oregon in Eugene, Oregon, preceding the Olympic Games in Los Angeles. The Congress was organized by the College of Human Development and Performance of the University of Oregon in collaboration with the cities of Eugene and Springfield. This was the first time in the history of the Congress that the event was organized by a group of private individuals, unaided by a federal government. The fact that the Congress was attended by more than 2,200 participants from more than 100 different nations is but one indication of its success.

The Congress program focused on the theme of Sport, Health, and Well-Being and was organized in three parts. The mornings of the eight-day event were devoted to disciplinary sessions, which brought together specialists in various subdisciplines of sport science such as sport medicine, biomechanics, sport psychology, sport sociology, and sport philosophy. For the first time in the Congress' history, these disciplinary sessions were sponsored by the national and international organizations representing the various subdisciplines. In the afternoons, the emphasis shifted toward interdisciplinary themes in which scholars and researchers from the subdisciplines attempted to contribute to crossdisciplinary understanding. In addition, three evenings were devoted to keynote addresses and presentations, broadly related to the theme of Sport, Health, and Well-Being.

In addition to the scientific programs, the Congress also featured a number of associative programs with topics determined by their sponsoring organizations. Well over 1,200 papers were presented in the various sessions of the Congress at large. It stands to reason, therefore, that publishing the proceedings of the event presented a major problem to the organizers. It was decided to limit proceedings initially to interdisciplinary sessions which drew substantial interest from Congress participants and attracted a critical number of high-quality presentations. Human Kinetics Publishers, Inc. of Champaign, Illinois, was selected to produce these proceedings. After considerable deliberation, the following interdisciplinary themes were selected for publication: Competitive Sport for Children and Youths; Human Genetics and Sport; Sport and Aging; Sport and Disabled Individuals; Sport and Elite Performers; Sport, Health, and Nutrition; and Sport and Politics. The 10-volume set published by Human Kinetics Publishers is rounded out by the disciplinary proceedings of Kinanthropometry, Sport Pedagogy, and the associative program on the Scientific Aspects of Dance.

Jan Broekhoff,
Michael J. Ellis, and
Dan G. Tripps

Series Editors

Preface

It is often said that "athletes are born and then they are made!" This rather simple statement implies, first, a natural endowment or genotypic potential for successful performance and second, the opportunity and environmental conditions to realize this potential. Thus, the interaction of the individual's genotype and his or her environment is emphasized. Athletes obviously are products of their genes *and* their environments, in addition to possessing some characteristics that covary with performance.

The determinants of sports performance are many, and sport scientists have expended considerable effort toward understanding phenotypic variation in size, physique, and body composition; metabolic powers and capacities; strength, speed, and skill; and cardiovascular adaptations relative to outstanding athletic performance. These efforts include not only the effects of training and practice, but also age- and sex-associated variation. Training and practice are important environmental components of the genotype-environment interaction; of course, response to these environmental factors varies with age and from individual to individual. However, relatively little research has been done on the contribution of genes to such human phenotypic characteristics. The same can be said for behavioral concomitants of athletic performance, for example, temperament. Although often attributed to social and cultural circumstances, there is evidence that a substantial portion of the variation in temperament has a genetic component (Buss & Plomin, 1975, 1984).[1]

This symposium and the papers that contribute to it represent a concerted effort by several human biologists, human geneticists, and sport scientists to

---

[1]Buss, A.H., & Plomin, R. (1975). *A temperament theory of personality development*. New York: Wiley.

Buss, A.H., & Plomin, R. (1984). *Temperament: Early developing personality traits*. Hillsdale, NJ: Erlbaum.

summarize aspects of the available literature and to present new observations on the genetic component of sport performance. Performance in the sport setting is indeed complex and not easy to study directly. To this end, the authors in this volume consider the genetic sources of variation in several contributory components of successful athletic performance.

Wilson summarizes the results of the Louisville Twin Study, focusing specifically on the physical growth (height and weight) of monozygotic and dizygotic twins from birth to 9 years of age. Malina presents an overview of motor development and motor performance, emphasizing that our efforts at understanding movement are largely oriented toward the product rather than the underlying processes and influencing factors. Data are reviewed for motor development during early childhood and then for performance in strength and motor tasks at older ages using observations on twins, siblings, and parents and their offspring. Bouchard summarizes the data from his laboratory and those of others concerning the heritability of maximal aerobic powers and aerobic capacities, as well as the genotype dependence of adaptive responses to training and other environmental variables. Schull presents a unique view of health-related fitness in terms of the genetic modulation of obesity, blood pressure, and plasma lipids. Although this perspective considers only a part of the spectrum of physical well-being, it is important to recognize the possible relationship of these conditions to the more commonly used indices of health-related fitness such as fatness, endurance, strength, and flexibility. Roberts attempts to synthesize the observations of the preceding four presentations, which dissect out certain contributory components of athletic excellence, and to place these observations in the context of the evidence for general control in such multifactorial traits.

These five papers are essentially synthetic, attempting to bring together a rather diverse range of information on several contributory components of athletic performance. The genetic influence on factors that may covary with performance are generally not considered in detail. These include, for example, somatic and sexual maturation during adolescence, physique and other morphological characteristics, perceptual characteristics, and behavioral traits.

The six contributed papers included in this volume are equally as diverse. Wolanski offers a brief summary of family studies (parent-offspring similarities) done in Poland over the past 20 years. The other contributed papers focus on the genotypic source of variance in several specific factors related to performance, such as cardiac size, muscle fiber composition, force-velocity relationships, anaerobic alactacid work capacity, and sensitivity to training.

Genetics has much to offer the sport sciences in terms of research on inheritance, genetic mechanisms, genotype-enviroment (i.e., training, learning), interactions, and so on. However, the two disciplines differ in their perspective, and the applications and implications of the observations of each often are not immediately apparent to the other. It is hoped that the contents of this volume will facilitate the dialogue among those in each discipline interested in athletic performance.

Robert M. Malina
Claude Bouchard

Editors

The 1984
Olympic Scientific
Congress
Proceedings
Volume 4

Sport
and
Human
Genetics

1

Twins: Genetic Influence on Growth

Ronald S. Wilson
UNIVERSITY OF LOUISVILLE SCHOOL OF MEDICINE
LOUISVILLE, KENTUCKY, USA

For over 20 years the Louisville Twin Study has been recruiting newborn twins for participation in a longitudinal study of growth and development. The twins have been measured periodically during childhood, and a set of growth standards for twins has been developed (Wilson, 1979).

The present paper appraises the degree of concordance in physical growth for monozygotic (MZ) twins and dizygotic (DZ) twins. The genotype is expected to play a substantial role in growth, leading to greater concordance for MZ twins, but this expectation is tempered by several other factors which would affect birth size and subsequent growth. The twins in each pair, whether MZ or DZ, share many prenatal influences and are delivered at the same gestational age, which should increase their similarity in birth size. For DZ twins, this might make them more concordant at birth than predicted on the basis of genetic overlap alone.

By contrast, about 70% of MZ twins are born with monochorionic placentas, and most of these placentas are subject to varying degrees of vascular anastomosis (Bulmer, 1970; Strong & Corney, 1967). If the anastomosis results in unequal nutrition being supplied to the twins, it will accentuate the within-pair differences in birth size. Naeye, Benirschke, Hagstrom, and Marcus (1966) have reported greater within-pair variability for monochorial twins, and on occasion a dramatic example of this transfusion syndrome may be found (e.g., Falkner, 1966).

This research has been supported in part by research grants from the National Institute of Child Health and Human Development (HD 14352), the Office of Child Health and Human Development (90-C-922), and the John D. and Catherine T. MacArthur Foundation. I am indebted to the many colleagues who have contributed so much to this program, including R. Arbegust, P. Gefert, M. Hinkle, J. Lechleiter, B. Moss, S. Nuss, and D. Sanders.

1

Longitudinal growth data on twins may be very instructive about the genetic contribution to individual growth curves. In pairs with large initial differences in birth size, successive measurements during infancy can demonstrate whether the differences are progressively offset or whether one twin continues to hold a significant advantage. Such measurements can also plot each child's distinctive growth curve, including episodes of spurt and lag, and further reveal whether twins follow the same pathways for spurt and lag.

The basic issues of interest can be phrased in the following questions:

1. Do genes appear to dictate the growth curve for each child, including spurts and lags?
2. Does the genetic template appear to pull a child back to its targeted growth curve if initially deflected by early deficit?
3. How closely do genetic replicates match each other for patterns of growth?
4. Comparatively, do DZ twins match each other more closely than other pairs of related zygotes from the same family?
5. For those occasional MZ pairs that do not match very closely, can the reasons be identified?

The salient issues may be illustrated by reference to the growth curves for two MZ pairs (Figure 1) and two DZ pairs (Figure 2). The height data for each twin has been plotted in standard-score form over successive ages from birth to 9 years of age.

The twins in the first MZ pair declined in parallel from above average at birth to below average at 6 months, then diverged at 3 years and displayed a consistent height difference thereafter. The second MZ pair, while more erratic during the first year, subsequently converged and remained virtually identical in height. This may be contrasted with the first DZ pair (Figure 2), who displayed larger and larger differences throughout childhood, or the second DZ pair, whose spurts and lags in growth curves were generally coordinated though the overall difference in height was never fully offset.

Sample and Procedure

The physical growth data are based on a sample of 952 twins, drawn from the entire socioeconomic range of the metropolitan Louisville area. The twins were measured at visits scheduled at 3, 6, 9, 12, 18, 24, 30, and 36 months, then annually thereafter. All twins were measured within ± 2 weeks of their birthdate (± 1 week in first year). There was slight attrition at older ages, due to a few families withdrawing from the study and the fact that not all the twins have yet reached their 9th birthday. No twins older than 3 months of age were added to the study, so the sample is truly longitudinal (Tanner, 1951) but has a few instances of missing data.

The methods of measurement have been described in detail elsewhere (Wilson, 1974). Weight was measured on a balance scale from 3 to 24 months, and on a platform scale at later ages. Birth weight was obtained from the twins' birth certificates or hospital records. Height was measured as recumbent length up to 24 months, and as standing height thereafter. Where available, measures of birth length were obtained from hospital records. In general, the birth length data were less complete and less precise than the length measures at subsequent ages.

The determination of zygosity for same-sex pairs was based on bloodtyping for 22 or more antigens (Wilson, 1980); if the twins were discordant for any antisera tests, they were classified as DZ. All concordant same-sex pairs were classified as MZ. Placental data were not available for most of these twins.

Results

For weight, the means and standard deviations were computed for each category of twins, and the results are presented in Table 1. The means were very close for MZ and DZ-SS (same sex) twins, and even the one sizable difference at 9 years was only marginally significant ($p = .06$). There appeared to be a slight but inconsequential weight bias of about 0.25 kg in favor of DZ-SS twins

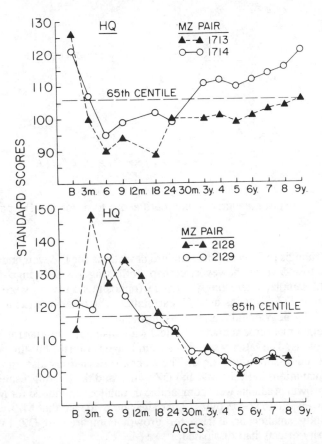

Figure 1. Height measures in standard-score form for two MZ pairs.

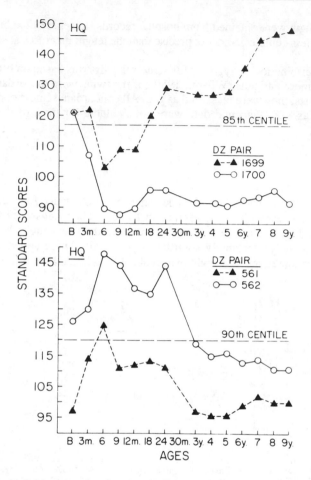

Figure 2. Height measures in standard-score form for two DZ pairs.

from 18 months to 8 years. The standard deviations for DZ weight did expand somewhat for 5 years, however, perhaps reflecting a wider range of zygotes in the DZ sample. Interestingly, DZ-OS (opposite sex) twins were the same in weight as DZ-SS twins until 6 years, after which DZ-OS twins averaged about 0.5 kg heavier (although nonsignificant).

The height measures shown in Table 2 were also quite similar for MZ twins and DZ twins. The latter were about 1 cm longer at birth, but this advantage was reduced at subsequent ages. The standard deviations also showed only minor fluctuations between MZ and DZ twins; and the overall results verified that the growth gradients were comparable throughout childhood for MZ twins, DZ-SS twins, and DZ-OS twins. There was no evidence that MZ twins were significantly smaller or at risk for growth compared to DZ twins after prematurity effects had dissipated.

Table 1. Weight (kg) of MZ twins and DZ twins from birth to 9 years

Ages	Mean weight			SD for weight			No. pairs		
	MZ	DZ-SS	DZ-OS	MZ	DZ-SS	DZ-OS	MZ	DZ-SS	DZ-OS
Birth	2.49	2.60	2.57	0.47	0.54	0.55	231	151	94
3 m.	5.29	5.41	5.40	0.78	0.76	0.73	176	123	67
6 m.	7.14	7.16	7.18	0.92	0.91	0.97	206	131	76
9 m.	8.34	8.47	8.38	1.06	1.03	1.09	199	127	78
12 m.	9.36	9.41	9.31	1.12	1.12	1.10	214	136	84
18 m.	10.57	10.82	10.62	1.22	1.28	1.35	210	129	78
24 m.	11.61	11.95	11.69	1.28	1.42	1.39	188	127	70
30 m.	12.75	13.07	12.99	1.49	1.53	1.60	109	81	57
3 y.	13.77	13.89	13.90	1.55	1.59	1.59	192	132	71
4 y.	15.65	15.88	15.76	1.73	1.96	2.05	184	121	69
5 y.	17.69	17.98	18.14	2.15	2.58	2.69	190	120	63
6 y.	20.10	20.31	20.78	2.73	3.37	3.22	175	111	58
7 y.	22.38	23.06	23.56	3.29	4.37	4.10	139	89	41
8 y.	25.84	26.05	26.55	4.21	5.54	5.13	156	88	57
9 y.	28.79	29.97	30.57	4.99	7.26	6.78	134	85	48

Note. MZ = monozygotic; DZ-SS = dizygotic–same sex; DZ-OS = dizygotic–opposite sex.

Height Recovery

On a samplewide basis, of course, twins are often premature in relation to singletons and they show a substantial deficit in birth size. The question is whether the prenatal stress of a twin pregnancy imposes a long-term deficit on growth, or whether there is substantial recovery. As mentioned earlier,

Table 2. Height (cm) of MZ twins and DZ twins from birth to 9 years

Ages	Mean height			SD for height		
	MZ	DZ-SS	DZ-OS	MZ	DZ-SS	DZ-OS
Birth	47.3	48.4	48.1	2.9	3.6	3.0
3 m.	57.6	58.4	58.0	2.8	3.0	3.0
6 m.	64.9	65.3	65.2	2.7	2.9	3.1
9 m.	69.4	70.2	69.7	2.8	2.8	3.4
12 m.	73.7	74.2	73.9	2.8	2.9	3.2
18 m.	80.2	80.8	80.2	3.1	3.4	3.4
24 m.	85.4	86.2	85.9	3.3	3.2	3.6
30 m.	89.0	90.0	89.9	3.4	3.5	3.8
3 y.	92.8	93.6	94.0	3.8	3.7	3.8
4 y.	100.3	101.1	101.2	4.1	4.2	4.1
5 y.	107.7	108.4	108.4	4.5	4.5	4.9
6 y.	114.5	115.0	115.0	4.7	5.0	5.1
7 y.	120.2	121.4	121.2	5.2	5.3	5.7
8 y.	126.5	127.1	127.0	5.9	5.8	5.9
9 y.	132.2	133.0	133.1	5.9	6.1	6.3

Note. Number of pairs the same as in Table 1 except for birth length, which was 168 (MZ), 92 (DZ-SS), and 71 (DZ-OS).

if the genetic template tended to restore each zygote toward its programmed growth curve, then the initial deficits due to prematurity might be offset during childhood.

The course of recovery may best be visualized by plotting a set of curves that compare the twins' height with that of singletons' from birth onward. The height measures were converted to standard-score form, and three representative centile values were selected to illustrate the course of recovery for small, medium, and large twins (10th, 50th, and 90th centiles, respectively). The curves are displayed in Figure 3.

The twins' height was sharply depressed in the first 3 months, but a gradual and steady recovery ultimately brought their height up to singleton norms by 8 years. The early effects of prematurity thus did not impose a long-standing liability on growth in this sample, and the twins ultimately reached parity with singletons for both height and weight (Wilson, 1979). Note that the extremes of the distribution (10th and 90th centiles) also reached the appropriate singleton level, so that the growth curves for twins were not compressed or skewed artificially by prematurity. In this perspective, the mechanisms regulating growth seem to be remarkably resilient in the face of prenatal stress, and twins as a group appeared to recovery fully to their targeted pathways of growth.

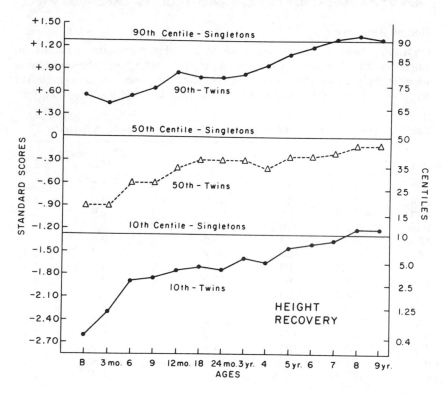

Figure 3. Height recovery for twins, birth to 9 years (from Wilson, 1979).

Zygosity and Growth

With this background on the complete sample, the question might be how closely MZ twins and DZ twins matched each other for height and weight throughout childhood. The pairs displayed in Figures 1 and 2 showed rather different patterns of concordance from age to age, and the question was how the concordance values might change when evaluated for all twins in each zygosity group.

Measures of concordance were obtained from within-pair (intraclass) correlations computed for each group, and the results are presented in Table 3. For weight, MZ twins were initially less concordant than DZ twins at birth but then steadily became more concordant over time as the DZ twins regressed to an intermediate level. The results were even more dramatic for height, with the MZ correlations rising to $R_{MZ} = 0.95$ at 8 years while the DZ correlations receded from $R_{DZ} = 0.77$ at birth to $R_{DZ} = 0.49$ at 8 years. The unusual features of twin pregnancies made MZ twins less alike at birth than expected, and DZ twins more alike (Price, 1950). But in the ensuing years, each zygosity group moved steadily toward a degree of concordance that reflected the genetic similarity within pairs. Height appeared to be more closely regulated by the genotype than weight, and it was less susceptible to variations in eating habits and food preferences that produced somewhat larger differences in weight.

A measure of dispersion within twin pairs is given by the within-pair variances, and the ratio of these variances for MZ twins and DZ twins indicates how much greater the dispersion of heights (or weights) was for DZ twins. The ratios are shown in the final two columns of Table 3, where they

Table 3. Twin within-pair correlations for weight and height

Ages	MZ	Weight DZ-SS	DZ-OS	MZ	Height DZ-SS	DZ-OS	Ratio DZ/MZ variances[a] Weight	Height
Birth	.64	.71	.67	.66	.77	.60	1.05	1.07
3 m.	.78	.66	.41	.77	.74	.61	1.44	1.30
6 m.	.82	.62	.48	.81	.70	.63	2.12	1.82
9 m.	.83	.55	.41	.83	.64	.51	2.52	2.25
12 m.	.89	.58	.42	.86	.69	.60	3.67	2.41
18 m.	.87	.54	.55	.89	.71	.64	3.88	3.31
24 m.	.88	.55	.51	.88	.59	.61	4.63	3.47
30 m.	.87	.55	.55	.93	.59	.62	3.61	5.72
3 y.	.89	.52	.55	.93	.59	.62	4.48	5.50
4 y.	.85	.50	.57	.94	.59	.58	4.36	7.09
5 y.	.86	.54	.62	.94	.57	.63	4.87	7.54
6 y.	.87	.57	.58	.94	.56	.58	5.26	8.28
7 y.	.88	.54	.54	.94	.51	.47	6.85	8.13
8 y.	.88	.54	.55	.95	.49	.52	6.47	9.15
9 y.	.88	.62	.51	.93	.49	.52	6.79	7.81

Note. MZ = monozygotic; DZ-SS = dizygotic–same sex; DZ-OS = dizygotic–opposite sex.
[a]Ratio of within-pair variances: DZ-SS/MZ.

progressed from unity at birth to values as high as 8 or more in the school years. Again, the values ran somewhat higher for height than for weight after 3 years, an indication that the dispersion of height differences was very limited in MZ pairs. The ratios also qualified as F-tests of within-pair variances, and from 9 months on the MZ variances were significantly smaller ($p < .001$). These results reinforced the conclusions from the correlation analysis and made it apparent that, for height especially, the similarities in twin growth moved inexorably toward a level commensurate with the number of genes shared in common.

Within-Pair Differences in Size

The distribution of weight differences within twin pairs gave a detailed picture of how large the actual differences were at each age for MZ and DZ twins. For illustration, the 50th- and 90th- centile values were selected to identify the average weight difference and the upper-range difference found for these twins. The centile values are presented in Table 4.

The median weight differences for MZ twins remained at 0.3 kg throughout the first 24 months, then gradually increased to 1.0 kg at 8 years. By contrast, the median weight difference for DZ twins increased steadily throughout childhood, reaching nearly 3.0 kg by 8 years. The upper-range differences followed a corresponding pattern, and in general the DZ weight differences exceeded the corresponding MZ differences by a factor of 2.0 to 3.0.

Interestingly, opposite-sex DZ pairs showed no greater differences in weight after 18 months than same-sex DZ pairs, although there were larger differences in the 3- to 18-month range. This dispersion within DZ-OS pairs reflected a pronounced sex difference in weight gain in the early months. After being comparable at birth, males gained weight more rapidly in the first 6 months, but

Table 4. Distribution of weight differences (kg) within twin pairs

Ages	50th centile difference			90th centile difference		
	MZ	DZ-SS	DZ-OS	MZ	DZ-SS	DZ-OS
Birth	.23	.27	.31	.65	.68	.68
3 m.	.31	.43	.54	.79	.96	1.16
6 m.	.31	.51	.73	.82	1.27	1.55
9 m.	.31	.63	.82	.83	1.44	1.87
12 m.	.31	.68	.79	.79	1.73	2.07
18 m.	.33	.77	.98	.91	1.92	2.04
24 m.	.32	.88	.99	.91	2.04	2.22
30 m.	.45	1.02	.79	1.25	2.60	2.49
3 y.	.37	.91	.79	1.25	2.60	2.38
4 y.	.57	1.36	1.25	1.59	3.06	2.94
5 y.	.57	1.59	1.36	1.70	3.85	4.08
6 y.	.79	1.81	1.59	2.27	4.42	4.98
7 y.	.79	2.38	2.49	2.38	6.11	6.68
8 y.	1.02	2.94	2.94	3.29	8.04	6.68
9 y.	1.22	3.28	3.17	4.31	9.06	10.30

Note. MZ = monozygotic; DZ-SS = dizygotic–same sex; DZ-OS = dizygotic–opposite sex.

this was progressively offset by catch-up growth for females around 24 months (Wilson, 1979). Thus, the median weight difference for DZ-OS pairs actually diminished in the third year and subsequently remained equal to or smaller than the corresponding DZ-SS differences.

Turning to height differences, the values for the 50th and 90th centiles are displayed in Table 5. For MZ twins, the median height difference declined after birth from 1.3 to 0.8 cm, then remained essentially constant to 6 years. It was remarkable that the average MZ height difference did not exceed 1.0 cm until 7 years, nor did the 90th-centile difference exceed 3.0 cm in the same period. This represented very tight regulation of height by the genotype, especially since the birth-size differences had been pronounced, and it clearly demonstrated that genetic replicates were pulled insistently toward a common pathway of growth. While a few pairs (to be discussed later) retained fairly large differences throughout childhood, they were the rare exceptions to the overall rule.

By contrast, DZ twins steadily increased in height differences over age, and after 24 months their median differences exceeded the MZ differences by a factor of 3 to 1. In fact, the 90th-centile MZ differences were smaller than the 50th-centile DZ differences from 5 years onward. Height disparity within a pair became an increasingly definite marker of dizygosity by school age, when the projected divergence within pairs was steadily played out in accordance with the genetic program. Note also that both opposite-sex and same-sex DZ twins coincided very closely for the distribution of height differences, except during the 9- to 24-month phase when sex differences in growth temporarily expanded the DZ-OS differences.

The trends over age may best be seen from curves that plot the median differences in height and weight throughout childhood. The plots for MZ twins

Table 5. Distribution of height differences (kg) within twin pairs

Ages	50th centile difference			90th centile difference		
	MZ	DZ-SS	DZ-OS	MZ	DZ-SS	DZ-OS
Birth	1.3	1.3	1.3	3.8	3.8	3.8
3 m.	1.0	1.5	1.7	3.1	3.5	4.2
6 m.	0.9	1.6	1.6	2.5	3.5	4.6
9 m.	0.9	1.4	2.2	2.5	3.9	5.0
12 m.	0.8	1.5	2.1	2.6	4.0	4.4
18 m.	0.8	1.8	2.1	2.2	4.4	4.3
24 m.	0.9	1.6	2.2	2.7	4.8	4.7
30 m.	0.9	2.3	2.2	1.9	5.0	4.9
3 y.	0.8	2.4	2.1	2.3	5.2	5.3
4 y.	0.9	2.4	2.3	2.3	5.9	6.0
5 y.	0.9	3.2	2.7	2.4	6.4	7.2
6 y.	1.0	3.9	3.2	2.5	7.5	6.9
7 y.	1.2	3.8	4.1	3.1	8.2	8.6
8 y.	1.2	4.3	4.1	3.0	9.0	9.1
9 y.	1.3	4.0	3.8	3.5	10.0	10.0

Note. MZ = monozygotic; DZ-SS = dizygotic–same sex; DZ-OS = dizygotic–opposite sex.

and DZ-SS twins are shown in Figure 4, with the height curves being read against the left ordinate and the weight curves against the right ordinate. The MZ curves are remarkably flat, especially for height. The median MZ differences simply did not increase very much throughout childhood, and this was during a time span when absolute height increased from 58 cm at 3 months to 132 cm at 9 years. By contrast, height differences in DZ-SS pairs showed a steep rise between 24 months and 6 years, with the median differences substantially exceeding the 90th-centile differences for MZ twins.

For weight, the MZ differences increased modestly after 5 years while the DZ differences showed a steady incremental gain year-to-year. Comparatively, weight differences were less tightly controlled by genetic factors than were height differences. There was greater weight dispersion in MZ pairs, and the 90th-centile difference did not fall below the 50th-centile DZ weight difference, as it had for height.

Birth Differences and Recovery

Figure 4 showed that MZ height differences were larger at birth than at subsequent ages; that is, for the full MZ sample the pairs pulled together more closely in height during the first year. On an individual basis, how were such changes manifested in the growth curves? The curves for two MZ pairs with large differences at birth are shown in Figure 5.

For pair #702-03, the large initial difference in birth length was rapidly offset during the first year, and the twins thereafter followed parallel trends in

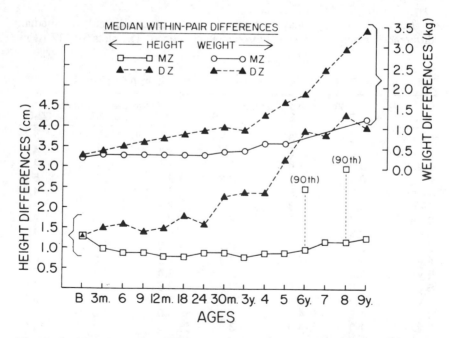

Figure 4. Median weight differences and height differences for MZ twins and DZ-SS twins.

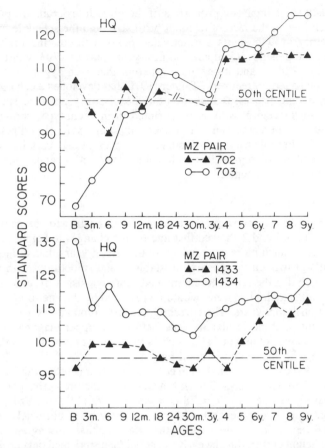

Figure 5. Growth curves (height) for two MZ pairs with large differences in birth length.

growth, with coordinated episodes of spurts and lags. Notably, the taller twin at 9 years was the one who had been smaller at birth, and while such a reversal was not common, it did illustrate that the smaller twin was not automatically at a long-term disadvantage relative to the larger twin.

The birth-length differences for the second pair (#1433-34) were offset less rapidly, and indeed the differences persisted up to school age. The growth spurt for #1433 between 4 and 7 years gradually pulled the height curve back into convergence with the co-twin and made it apparent that some match-to-model process still operated to restore the smaller twin to its targeted growth curve.[1]

---

[1]A fascinating sidelight: #1433 was hospitalized at 4 years for internal bleeding from the colon, a problem that recurred for 2 years and led to exploratory surgery at age 6. Nothing was found, and by 7 years the bleeding had stopped. Thus, the catch-up growth occurred during a period when #1433 was otherwise debilitated by internal bleeding.

Obviously some intrinisic properties of the growth program are preserved as latent entities in the face of unequal birth size, and the schedule by which compensatory growth spurts are activated poses a fascinating problem for developmental genetics. Compare the first-year spurt of #703 with the much later spurt of #1433, and it becomes apparent that each zygote had its own idiosyncratic pattern of recovery. Tanner (1970) has developed a detailed theory of self-regulating processes that monitor the disparity between suppressed growth and projected growth as blueprinted in the genotype, showing how these regulatory processes serve to offset temporary deficits and restore the child to his or her intrinsic growth curve. The present data certainly confirm this theory and further suggest that each child's "sizostat" operates on a distinctive chronological schedule.

Age-to-Age Stability

In addition to showing convergence among twins, the growth curves in Figure 5 and in Figures 1 and 2 showed that age-to-age changes in height, or spurts and lags, were much more pronounced in the first 2 years than at later ages. Each twin's growth curve appeared to stabilize and smooth out with increasing age—such that the HQ scores remained more consistent from age to age.

For the full sample, this trend would be revealed by the age-to-age correlations from birth to 9 years. The correlations were computed between all ages for MZ twins, and the results are presented in the upper diagonal matrix of Table 6. The final column of Table 6 gives the correlation of the twin's height at each age with the mother's adult height.

The entries showed that birth length correlated only modestly with later height ($r = .46$ at 12 mos, $r = .26$ at 3 yrs); but by 1 year the predictive correlations had increased considerably ($r = .78$ at 3 yrs, $r = .72$ at 9 yrs). From 3 years on, the age-to-age consistency for height was very high, ultimately reaching $r = .99$ at 8 years. The results verified the inference that most age-to-age fluctuations in height occurred in the early years but then smoothed out considerably by school age. In genetic analysis, such age-to-age correlations are often referred to as phenotypic correlations.

The entries along the diagonal are the MZ within-pair correlations for height at each age, and they also steadily increased to terminal values of 0.93-0.94 at 9 years. From 5 years onward they remained slightly below the age-to-age correlations, indicating that MZ twins did not match each other quite as closely as each twin matched him- or herself year-to-year. As noted earlier, while most MZ pairs converged on a common developmental pathway, a few retained differences in height that persisted to 9 years. Thus, some discrepancies in height must be attributed to factors that had an unequal effect upon the shared genotype of these MZ twins.

Cross-Correlations for Height

The entries below the diagonal are the cross-correlations for MZ twins; that is, Twin A's height at 3 months correlated with Twin B's height at 6 months, and so on for all combinations. The correlations jointly reflected the concordance within MZ pairs plus the stability of the height measures across ages. As the entries showed, the cross-correlations became very high at the later

Table 6. MZ twins: Intercorrelations and cross-correlations for height

Ages	B	3 m.	6	9	12	18	24	30 m.	3 y.	4	5	6	7	8	9 y.	Maternal Height
B	(.68)	.64	.52	.48	.46	.36	.38	.40	.26	.24	.26	.28	.26	.28	.27	.06
3 m.	.50	(.76)	.82	.80	.76	.66	.58	.58	.54	.49	.50	.49	.50	.47	.44	.18
6 m.	.40	.68	(.80)	.88	.85	.78	.72	.73	.68	.62	.60	.56	.58	.58	.55	.23
9 m.	.40	.66	.76	(.83)	.90	.85	.80	.80	.74	.68	.70	.66	.66	.68	.66	.29
12 m.	.38	.66	.76	.81	(.86)	.88	.85	.81	.78	.73	.76	.73	.75	.74	.72	.32
18 m.	.30	.56	.70	.77	.82	(.89)	.92	.90	.84	.80	.80	.78	.78	.77	.73	.38
24 m.	.32	.50	.63	.72	.78	.86	(.89)	.96	.91	.86	.85	.80	.82	.81	.77	.47
30 m.	.35	.52	.66	.74	.75	.85	.89	(.93)	.97	.94	.91	.86	.88	.86	.80	.50
3 y.	.22	.46	.60	.67	.72	.76	.84	.91	(.93)	.94	.92	.88	.88	.86	.82	.48
4 y.	.20	.42	.54	.60	.67	.74	.80	.88	.88	(.93)	.95	.92	.92	.90	.86	.51
5 y.	.21	.44	.54	.62	.70	.74	.79	.86	.86	.89	(.94)	.97	.94	.94	.90	.46
6 y.	.24	.43	.50	.59	.68	.72	.74	.82	.83	.86	.92	(.94)	.98	.98	.94	.42
7 y.	.22	.43	.52	.60	.70	.73	.78	.83	.83	.86	.90	.92	(.93)	.99	.98	.44
8 y.	.24	.42	.53	.63	.70	.72	.77	.82	.82	.84	.90	.93	.94	(.94)	.99	.44
9 y.	.21	.36	.48	.58	.68	.68	.72	.75	.76	.79	.86	.90	.90	.93	(.93)	.39

Note. Diagonal entries are MZ within-pair correlations at each age; upper diagonal entries are age-to-age correlations for all MZ twins; lower diagonal entries are cross-correlations for Twins A and B in MZ pairs. Final column is correlation of mother's adult height with twins' height. MZ = monozygotic.

ages and virtually duplicated the within-pair correlations in the diagonal. If the height measures across two ages were very stable (e.g., $r = .99$), then the cross-correlations for MZ twins would be virtually identical to the within-pair correlations at each age. The implications of these results will be considered later.

Finally, the correlations of twins' height with maternal height are shown in the last column of Table 6; they moved from zero-order at birth to approximately $r = .50$ by 3 years, then receded very slightly in subsequent years. It appeared that as growth proceeded, the infants of taller mothers tended to become taller than average whereas the infants of smaller mothers remained below average in height. While one might anticipate such a relationship when the twins reached adult height, it was remarkable to find a correlation of $r = .50$ at only 3 years. Obviously, familial similarity in relative height became evident at an early age.

DZ Correlations

Age-to-age correlations and cross-correlations were also computed for DZ twins, and the results are presented in Table 7. The age-to-age correlations in the upper diagonal matrix were a virtual duplicate of the corresponding MZ correlations, so the stability in height followed the same pattern in both zygosity groups. The cross-correlations in the lower diagonal matrix were markedly different, however, especially at the later ages. They were limited by the DZ within-pair correlations shown in the diagonal, and as these regressed to $r = 0.49$, the cross-correlations drifted toward the same level.

At 5 years, Twin A predicted Twin B's current height moderately well ($r = 0.55$), but it was predicted almost as well by Twin A's height at any previous age back to 30 months ($r = 0.46$ to 0.50). Thus, while each twin became more consistent with himself or herself over ages, the height similarity within DZ pairs progressively declined to an intermediate level. As a consequence, the prediction of twin B's height was about as well accomplished by any height measure on Twin A from 30 months onward as by the concurrent measure. This may be contrasted with the steadily increasing cross-correlations for MZ twins, where predictability improved with age.

The final column of Table 7 presents the correlations between maternal height and that of the DZ twins. The correlations closely paralleled the values for MZ twins. All twins moved progressively toward an intermediate degree of similarity in relative height with their mothers, although in absolute terms the mothers averaged 70 cm taller than their 3-year-old offspring.

Plot of Concordance Relationships

The contrasting relationships among these groups of related zygotes may be visualized from curves that displayed the trends over ages. These curves are presented in Figure 6 and show the phenotypic correlations at adjacent ages (birth to 3 months...8 years to 9 years), the MZ and DZ within-pair correlations, and the mother-child correlations over ages.

The age-to-age correlations started at $r = 0.66$ and moved steadily upward until stabilizing in the upper 0.90s. The MZ within-pair correlations followed a very similar trend until stabilizing in the mid-0.90s after 5 years, slightly

Table 7. DZ same-sex twins: Intercorrelations and cross-correlations for height

Ages	B	3 m.	6	9	12	18	24	30 m.	3 y.	4	5	6	7	8	9 y.	Maternal Height
B	(.79)	.74	.50	.54	.44	.40	.38	.30	.36	.37	.40	.46	.48	.42	.35	.03
3 m.	.63	(.75)	.84	.80	.74	.64	.54	.54	.47	.40	.44	.51	.50	.50	.50	.18
6 m.	.42	.64	(.69)	.84	.83	.75	.64	.66	.53	.48	.52	.56	.56	.54	.56	.17
9 m.	.44	.60	.59	(.63)	.89	.82	.76	.70	.64	.58	.60	.64	.62	.62	.60	.26
12 m.	.36	.54	.60	.61	(.68)	.89	.82	.80	.70	.68	.68	.70	.66	.66	.65	.28
18 m.	.30	.46	.55	.55	.62	(.70)	.90	.86	.80	.77	.76	.78	.74	.73	.72	.32
24 m.	.30	.34	.41	.48	.50	.53	(.57)	.92	.86	.83	.81	.82	.78	.78	.74	.40
30 m.	.23	.34	.42	.42	.50	.53	.48	(.55)	.95	.92	.90	.89	.86	.86	.82	.46
3 y.	.30	.26	.30	.37	.41	.50	.49	.55	(.59)	.95	.90	.93	.90	.87	.84	.44
4 y.	.30	.22	.26	.34	.39	.48	.44	.48	.54	(.59)	.95	.97	.96	.90	.88	.42
5 y.	.32	.26	.30	.35	.41	.44	.42	.50	.49	.53	(.55)	.55	.98	.94	.92	.44
6 y.	.38	.32	.34	.40	.42	.47	.44	.52	.50	.52	.53	(.55)	.50	.98	.96	.48
7 y.	.43	.32	.34	.39	.41	.44	.40	.49	.46	.46	.48	.50	(.49)	.99	.98	.40
8 y.	.36	.32	.29	.36	.36	.42	.37	.48	.46	.46	.46	.50	.48	(.48)	.99	.50
9 y.	.27	.32	.34	.34	.38	.39	.36	.48	.44	.46	.48	.50	.43	.46	(.49)	.46

Note. Diagonal entries are DZ within-pair correlations at each age; upper diagonal entries are age-to-age correlations for all DZ twins; lower diagonal entries are cross-correlations for Twins A and B in DZ pairs. Final column is correlation of mother's adult height with twins' height; DZ = dizygotic.

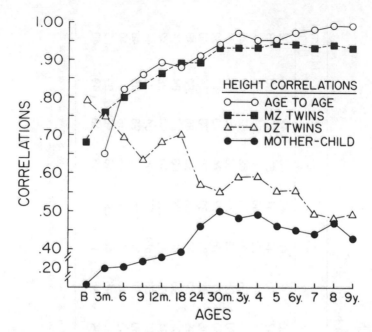

Figure 6. Height correlations for MZ twins, DZ twins, and mother-child sets; and for each child with itself, age to age.

below the phenotypic correlations. The DZ correlations, by contrast, began at a higher level ($r = 0.79$) but then steadily declined to $r = 0.49$, while the mother-child correlations progressed from zero to the mid-0.40s. In terms of concordance for height, the trends leaned strongly toward the number of genes shared in common, approaching unity for MZ twins and approximating 0.50 for DZ twins and mother-child pairs.

The directional trends over age were particularly persuasive in showing the steady march of height measures toward their targeted endpoints, as guided by an intrinsic groundplan. It was remarkable that the similar experiences shared by DZ twins in the preschool years did not sustain their concordance significantly above mother-child concordance, since all possible effects of age differences and secular trends would bear upon the latter relationships. And when these two-zygote curves were contrasted with the steady gain in MZ concordance, it seemed abundantly clear that the genotype dictated the growth in height over a wide range of modifying conditions.

Estimating Genetic Covariance

In longitudinal growth data such as these, it is often informative to estimate the common genetic influences on height measures obtained at any two ages. To the extent that there was continuity in relative height over age, it would reflect the ongoing contributions of a particular complement of genes, active throughout the period. But if there were spurts or lags that shifted relative height from one age to the next, it would reflect the action of distinctive gene sets that switched on and off during the period. Intermittent environmental influences on growth would also be combined with the latter effect, while sustained environmental influences would combine with the ongoing genetic effect.

Data on twins make it possible to estimate the common genetic influence on growth via path analysis (Plomin & DeFries, 1981). The phenotypic correlations can be decomposed into a component representing continuing genetic influence, termed the genetic covariance, plus a component representing environmental stability. The calculations make use of the cross-correlations for MZ twins and DZ twins as presented in Tables 6 and 7. The genetic covariance between any two ages is calculated from the cross-correlations as 2 $(R_{MZ_{i.j.}} - R_{DZ_{i.j.}})$. These have been computed for all sets of cross-correlations and are presented in the upper diagonal matrix of Table 8. The entries represent the inferred contribution of the same genes to the longitudinal stability of height measures over age.

The entries ranged from low-order and erratic at birth to very large and consistent at the later ages. From 12 months onward, the genetic contribution to longitudinal stability remained reasonably consistent for each succeeding age, but the size of the contribution increased as the predictor age increased. At 8 and 9 years of age, the genetic covariance approached its upper limit as the MZ/DZ cross-correlations approached their theoretical levels ($r = 0.93$ and 0.46, respectively). Nearly all of the longitudinal stability in height measures at these later ages could be attributed to the continuing action of the same gene complement.

To what extent was longitudinal stability in height not accounted for by the same genes? This may be computed by subtracting the genetic covariance from the phenotypic correlation, the result of which identifies the residual stability in height arising from other sources. For clarity of illustration, the figures have been computed only for adjacent ages and are shown in the single vector of lower-diagonal entries in Table 8.

The nongenetic value was very high initially (recall that the DZ correlation actually exceeded the MZ correlation for this period), but at each succeeding age the values dropped until reaching $r_e = .05$. In the first 18 months, a great deal of the longitudinal stability between ages was accounted for by nongenetic factors, reflecting the prenatal influences that had enhanced DZ similarity and reduced MZ similarity. But as these influences waned and as each twin moved steadily onto his or her own growth curve, the MZ and DZ correlations moved toward their expected levels (cf. Figure 6) and the genetic covariance increased.

Time-Linked Gene Action

The opposite side of genetic covariance is the concept of age-linked gene action that would affect height at one age but not the next. To the extent that phenotypic correlations fell below 1.00, it was indicative of time-limited genetic effects plus intermittent environmental effects that produced uneven growth spurts. At closely spaced ages this effect would be negligible except in the first year, but for more distant ages it would become a significant factor. Two of the twins in Figure 2 showed dramatic changes in relative height between early and later ages; and on a sample-wide basis the age-to-age correlation between height at 1 year and at 9 years was 0.65. The diminished age-to-age

Table 8. Genetic covariances for continuity in height across ages

Ages	B	3m.	6	9	12	18	24	30 m.	3 y.	4	5	6	7	8	9 y.
								Ages							
B															
3 m.	.95	-.26	-.04	-.08	.04	.00	.04	.24	-.16	-.20	-.22	-.28	-.42	-.24	-.12
6 m.		.75	.08	.12	.24	.20	.32	.36	.40	.40	.36	.22	.22	.20	.08
9 m.			.52	.34	.32	.30	.44	.48	.60	.56	.48	.32	.36	.48	.28
12 m.				.50	.40	.44	.48	.64	.60	.56	.54	.38	.42	.54	.48
18 m.					.48	.40	.56	.50	.62	.56	.58	.52	.58	.48	.60
24 m.						.25	.66	.64	.52	.52	.60	.50	.58	.60	.58
30 m.							.12	.82	.70	.72	.74	.60	.76	.80	.72
3 y.								.24	.72	.80	.72	.60	.68	.68	.54
4 y.									.26	.68	.74	.66	.74	.72	.64
5 y.										.23	.72	.68	.80	.76	.66
6 y.											.19	.78	.84	.88	.76
7 y.												.14	.84	.86	.80
8 y.													.07	.92	.94
9 y.														.05	.94

Note. Genetic covariances obtained from cross-correlations by formula $2(R_{MZi,j} - R_{DZi,j})$. Single-lower-diagonal entries obtained from $[r_{ij} - 2(R_{MZi,j} - R_{DZi,j})]$ at each age. See text for explanation.

correlations illustrated the extent of time-linked gene action in shaping the course of growth for each child.

Enduring MZ Differences

As a final point, the last lower-diagonal entry of 0.05 in Table 8 represented the degree of height stability not accounted for by genetic covariance. In essence it reflected the reliable differences in height among MZ pairs that had not been offset by 9 years of age. It brought back the question posed earlier about the origin of such differences; for illustration, two MZ pairs with divergent height curves are shown in Figure 7.

The top pair (#824-25) represented the most dramatic and long-standing difference in height among all MZ twins, while the bottom pair diverged substantially after the first year but then reconverged by school age. Note that the

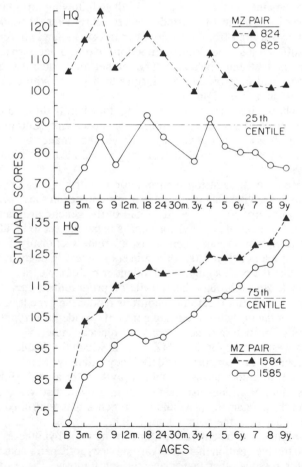

Figure 7. Growth curves (height) for two MZ pairs with long-sustained differences in height.

directional changes in the curves were well coordinated and in phase for both sets of twins. The differences were in overall level rather than the ebb and flow of the growth curves.

Such differences are sometimes ascribed to placental anastomosis, in which the blood supply of one twin's placenta is transfused to the other twin's placenta by an arterio-venous shunt. Anastomosis occurs only in monochorionic placentas, however, and both of these pairs were delivered with dichorionic placentas, which would seem to rule out any vascular interchange.

Turning to the health histories, twin #825 had remained in the hospital for 1 month following birth, and both twins suffered from viruses, bronchial conditions, and sore throats in 2nd and 3rd years—the larger twin more severely so than the smaller. At 4 years, #825 had periods of rising and falling temperatures; and at 16 years the mother reported that the twin had been diagnosed as having Osgood-Schlatter disease, which was "caught and repaired." No other health problems were reported.

Twin #1585 remained in the hospital for 11 days following birth, and from 3 months onward was subject to periodic fevers nearly every 2 weeks. At 3 years, both twins suffered from tonsillitis, ear infections, and allergies; at 5 years #1585 suffered a serious kidney infection, and at 6 years she was put on Atafax for "high-strung" behavior. The mother terminated the drug at 7 years. Both girls had outgrown their allergies at 8 years, with their health reported as good and free of problems.

Although the health histories were interesting, they hardly differed from the usual chronicle of childhood diseases found for most twins. Certainly there was no striking illness or defect that had afflicted the smaller twin. Growth hormone deficiency might be suspected, although no clinical test was reported by the mothers.

In the absence of the transfusion syndrome or any striking health problem, perhaps we could speculate about the growth differences in these MZ pairs. Given the extraordinary complexity and detail of the genetic program which must be transcribed at each cell division, perhaps factors yet to be identified affect the transcription in early cell division, and fractionally lower the replicability coefficient below 1.00 for a subset of genes. Recent advances in genetics show that the code is highly redundant, which may explain how the course of growth can be brought back to its preprogrammed curve in spite of induced deflections. But if certain conditions render the gene transcription less than exact for the growth-influencing genes of two identical zygotes, then the parameters of growth may not be precise replicas either.

Perhaps this was the case for three MZ pairs in which the smaller twin suffered from idiopathic growth hormone deficiency (Lindsay, MacGillivray, & Voorhees, 1982); in milder terms it might apply to #825 as well. For #1585 the effect was more of a long-sustained delay in growth that was eventually overcome, and the primary focus would be on genes that controlled maturational phases rather than the twin's adult height.

Other possibilities may be entertained, but for now the hypotheses are less important than the recognition that some rare monozygotic pairs do not match very closely for reasons other than placental anastomosis. These pairs cast in sharp relief some central issues for developmental genetics which have far-reaching implications for the regulation of growth.

References

Bulmer, M.G. (1970). *The biology of twinning in man*. Oxford: Clarendon Press.

Falkner, F. (1966). General considerations in human development. In F. Falkner (Ed.), *Human development* (pp. 10-39). Philadelphia: Saunders.

Lindsay, A.N., MacGillivray, M.H., & Voorhees, M.L. (1982). Growth hormone deficiency in twins: Three cases with normal co-twins. *Pediatrics, 69*, 486-488.

Naeye, R.L., Benirschke, K., Hagstrom, J.W.C., & Marcus, C.C. (1966). Intrauterine growth of twins as estimated from liveborn birthweight data. *Pediatrics, 37*, 409-416.

Plomin, R., & DeFries, J.D. (1981). Multivariate behavioral genetics and development: Twin studies. In L. Gedda, P. Parisi, & W.E. Nance (Eds.), *Twin research 3, part B* (pp. 25-33). New York: Alan R. Liss.

Price, B. (1950). Primary biases in twin studies: A review of prenatal and natal difference-producing factors in monozygotic pairs. *American Journal of Human Genetics, 2*, 292-322.

Strong, S.J., & Corney, G. (1967). *The placenta in twin pregnancy*. Oxford: Pergamon.

Tanner, J.M. (1951). Some notes on the reporting of growth data. *Human Biology, 23*, 93-159.

Tanner, J.M. (1970). Physical growth. In P.H. Mussen (Ed.), *Carmichael's manual of child psychology* (Vol. 1, pp. 77-155). New York: Wiley.

Wilson, R.S. (1974). Growth standards for twins from birth to four years. *Annals of Human Biology, 1*, 175-188.

Wilson, R.S. (1979). Twin growth: Initial deficit, recovery, and trends in concordance from birth to nine years. *Annals of Human Biology, 6*, 205-220.

Wilson, R.S. (1980). Bloodtyping and twin zygosity: Reassessment and extension. *Acta Geneticae Medicae et Gemellologiae, 29*, 103-120.

2

Genetics of Motor Development and Performance

Robert M. Malina
UNIVERSITY OF TEXAS
AUSTIN, TEXAS, USA

Motor development refers to the process through which a child acquires movement patterns and skills. It is a process of continuous modification based upon the interaction of the genotypically controlled rate of neuromuscular maturation, residual effects of prior experiences, and the new motor experience. In contrast, motor performance is viewed most often in the context of tasks that are performed under specified conditions and that are amenable to precise measurement (Malina, 1982). During the first 6 or 7 years of life, children develop a series of fundamental movement patterns. There is considerable variation, however. Most children are able to perform the basic movement patterns by 6 or 7 years of age, but their levels of skill vary considerably. Skill implies accuracy as well as economy and precision of performance and shows a high degree of inter-individual variability.

Like most biological traits, many motor characteristics show a moderate heritability. This review attempts to provide an overview of studies that deal with estimates of the genotypic contribution to motor development and performance. In addition, several factors associated with motor performance, perhaps covariates of performance, are also briefly considered. When possible, this review treats results from a number of studies as a group. In many instances this is not possible, however, given the different tasks used and different methods of estimating heritabilities.

Motor development and performance can be viewed in terms of the physiological, biomechanical, and psychological processes that underlie the performance of a motor task—for example, changing lever arms, angle of projection, enzyme activities in muscle tissue, impulse transmission, and so on. Such processes are quite difficult to study in the performance setting, and most emphasis is placed upon biomechanical aspects using cinematographic analyses.

The product of performance refers to the outcome of the movement activity, that is, the distance jumped, the time elapsed, errors per second, and so on. Most of the data on motor development and performance deal with the products of movement, and this is also true of studies that attempt to estimate the genotypic contribution to variance in motor tasks.

The available data relevant to the genetics of motor development and performance are based primarily on the twin model, which has limitations such as inequality of environmental covariance of monozygotic (MZ) and dizygotic (DZ) twins, small sample sizes, differential effects of age and sex according to twin types, and differences in means and variances between twin samples (Bouchard & Malina, 1983a). One can ask whether MZ twins are more similar due to genetic similarity or to environmental pressure for similarity. Or, how important is the role of mutual imitation in the motor development of twins? Hence, estimates of the genetic contribution to motor development and performance must also include other genetic relatives such as siblings, parents and offspring, and cousins. There are relatively few studies of motor characteristics of such relatives.

The effects of environmental conditions on the similarities between genetic relatives in motor performance is not often considered. One of the more significant environmental factors that should be controlled is the degree of habitual physical activity, which quite often occurs in the sports setting. Specific motor skills are practiced in such settings, and many tests of motor performance are based upon these skills, for example, the ball throw for distance.

Variation in motor performance is partly a function of the reliability of measurements. Performance is ordinarily measured under specified test conditions, and individuals vary from trial to trial and from day to day. Correlations between repeated measurements of the same individual are normally used to estimate a test's reliability or repeatability. Reliability of measurement should be considered in calculating heritability estimates, but this is not always done.

The data on genetics of motor development and performance are most often reported in terms of differences between twins, correlations, and heritabilities, although other estimates are occasionally used. All are sample or population specific, and it is reasonable to assume that genetic-environment interactions could affect familial correlations and heritability estimates. Other methodological concerns in evaluating the genetics of motor development and performance have been reviewed in more detail elsewhere (Bouchard & Malina, 1983a).

Motor Development

Studies during the first 2 years of life commonly utilize scales of motor development (e.g., the Bayley or Gesell scales of infant development). Results of twin studies indicate greater similarity within MZ pairs than within DZ pairs, that is, significantly greater within-pair differences for DZ twins (Freedman, 1974;

Freedman & Keller, 1963; Wilson & Harpring, 1972). Within-pair correlations for the motor development of twins are shown in Table 1. Note that the degree of concordance within MZ pairs varied with age, especially between 12 and 18 months. This decrease in concordance was attributed to difficulty in testing youngsters at these ages because of tantrums, crying, uncooperativeness, and so forth (Wilson & Harpring, 1972).

MZ and DZ twins do not differ significantly from each other in motor achievements on the developmental scales, nor does the firstborn twin differ as a rule from the second born in early motor development (Dales, 1969). However, twins do tend to show a consistent lag in motor development compared to singletons. This is shown in Figure 1 for the combined gross motor and adaptive behavior scores of the Gesell Developmental Schedule (Dales,

Table 1. Within-pair correlations for scores on the motor scale of the Bayley Scales of Infant Development

Age, months	MZ (n = 71-93)	DZ (n = 77-91)
3	.50	.41
6	.87*	.75
9	.84*	.61
12	.75	.63
18	.70	.77

Note. Data compiled from Wilson and Harpring (1972).
*p < .05

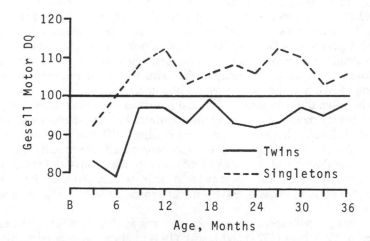

Figure 1. Gesell motor developmental quotients for twins and singletons (drawn from the data of Dales, 1969).

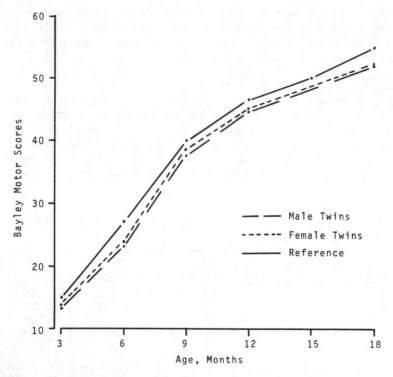

Figure 2. Bayley motor scores for male and female twins compared to reference data (drawn from the data of Wilson & Harpring, 1972, and Bayley, 1969).

1969), and in Figure 2 for the Bayley Motor Scale (Wilson & Harpring, 1972). The adaptive component of the Gesell schedule is largely based on fine motor items. Using the gross and fine motor components of Developmental Indicators for the Assessment of Learning (DIAL test), Cook and Broadhead (1984) also noted significantly poorer motor achievements in 2.5- to 5.5-year-old twins compared to singletons. Zygosity of the twins was not considered, so comparison of MZ and DZ twins was not possible.

Twins often differ in size at birth and are often born prematurely. Eight-month Bayley motor scores for twins dissimilar in size are shown in Table 2. The evidence indicated no significant intrapair differences. However, all subsamples of twins had lower mean motor scores than race-specific reference data for the Bayley scales (Bayley, 1965). In a similar analysis of within-pair differences of heavier and lighter twins at birth, Wilson and Harpring (1972) also noted no differences in Bayley motor scores between 3 and 18 months of age.

Data dealing with specific movement patterns are less extensive. The early studies of Verschuer (1927) and Bossik (1934) indicate greater concordance among MZ twins (67% and 69%) than among DZ twins (35% and 30%) for first efforts at walking. However, the timing of sitting behavior was almost

Table 2. Motor development of monozygotic co-twins with dissimilar birth weights

Percent difference birth weight	n pairs	Mean 8-month motor score		Motor score differences		
		Heavier	Lighter	Mean	S_M	t
White: < 15%	32	31.9	31.2	0.43	0.58	0.75
≥ 15%	16	28.3	27.5	0.75	0.65	1.15
Black: < 15%	51	29.9	29.4	0.45	0.40	1.13
≥ 15%	16	29.6	29.6	0.06	0.82	0.07

Note. Data compiled from Fujikura and Froehlich (1974). Mean 8-month motor scores for reference data: Black, 35.7; White, 34.4 (Bayley, 1965).

equally concordant among MZ and DZ twins (82% and 76%, respectively). The development of other movement patterns during early childhood such as crawling, stair climbing, and running generally indicates significantly greater concordance among MZ twins (Gesell, 1954), but much of the data are based upon comparisons of single pairs of twins (Gesell & Thompson, 1929; Gifford, Murawski, Brazelton, & Young, 1966; McGraw, 1935; Smith, 1976). Early motor development data for biological siblings are not available.

Motor Performance

Studies of motor performance emphasize the outcome of performance in specific tasks, primarily in school-age children and youth. Although a distinction is made between fine and gross motor skills, it should be recognized that many motor tasks incorporate both fine and gross motor control. Furthermore, muscular strength, often considered as a separate aspect of performance, is an essential component of especially gross motor performances.

Muscular Strength

Strength is an expression of muscular force, an individual's capacity to exert force against some external resistance. Three kinds of strength measurements are commonly indicated: (a) static or isometric strength is force exerted without any change in muscle length; (b) dynamic strength is the force generated by shortening (concentric contraction) or lengthening (eccentric contraction) the muscles (Asmussen, 1968); and (c) explosive strength, which is more variably defined, is an index of muscular power—the ability of muscles to release maximal force in the shortest possible time (Clarke, 1967). The genetic contribution to individual differences, primarily in static and dynamic strength, are considered in this section. Indicators of explosive strength, for example, the standing long jump, vertical jump, or ball throw for distance, are considered subsequently in the section dealing with the performance of runs, jumps, and throws.

Twin Studies

Data on the genotypic contribution to muscular strength are diverse in kinds of tests used and estimates of genetic influences. Three studies from Japan in the late 1940s and early 1950s included several strength measurements among a variety of motor fitness tests. Kimura (1956) considered dynamometric back strength in primary and junior high school twins (grades 1 through 9) from Osaka, while Ishidoya (1957) considered dynamometric measures of grip, shoulder, and back strength in twins 6 to 16 years of age from Sendai. Both studies used Verschuer's (1925) method of percentage deviation within twin pairs, although this method is not presently used.[1] Mizuno (1956) described similarities in several strength measurements in twins and unrelated pairs of boys of junior and senior high school age (grades 7 through 12) from Tokyo. The analysis was based on partial correlations and percentage differences within pairs. Types of twins, numbers of pairs, and comparisons of the three studies are presented in Table 3. Kimura's (1956) data showed no significant difference between MZ and DZ twins of each sex, respectively, in back strength.

Male MZ twins, however, showed a significant reduction in the percentage deviation with age, while male DZ twins and female MZ and DZ twins did not. On the other hand, Ishidoya (1957) reported significantly smaller percentage deviations in MZ twins than in like-sex DZ twins for one dynamic (push-ups) and three static strength tests. Results for the static strength tests were similar when adjusted for size variation. Although numbers were small, the results were essentially the same for comparisons between twins within specific age groups. Ishidoya (1957) also calculated an "heredity index" ("Erbkraft") and an index of environmental influence. The estimated genetic contribution to back and shoulder strength was about 60%, while that to grip strength and push-up performance was about 40%.

Mizuno (1956) used male and female MZ twins and pairs of grade-matched unrelated males. Male MZ twins had smaller percentage differences within pairs than did unrelated pairs of boys in dynamic (pull-ups) and three static

Table 3. Japanese studies of strength in twins: Mean percent deviation/difference within pairs

Strength tests	Ishidoya (1957) Like-sex MZ (57)[a]	DZ (44)	Kimura (1956) Males MZ (138)	DZ (53)	Females MZ (133)	DZ (78)	Mizuno (1956) Male MZ (15)	Unrelated pairs (27-29)
Grip	4.54	8.57					R 11.24	15.11
							L 7.47	17.25
Shoulder	5.46	12.83						
Back	4.01	9.84	5.7	5.2	6.1	6.2	17.31	23.18
Push-ups	10.22	17.22						
Pull-ups							63.93	77.40

[a]Numbers in parentheses are the number of twin pairs.

[1]Hoshi, Takahashi, Ashizawa, and Kouchi (1980) have described a revision of this method, the Revised Percent Deviation method.

strength tests. Partial correlations controlling for age tended to be higher in 15 pairs of male MZ twins than in 12 pairs of female MZ twins for the four strength tests, while those for the grade-matched pairs of unrelated boys were uniformly low.

Heritability estimates for measures of static and dynamic strength in twins come largely from Eastern Europe. Skład's (1973) study of Polish twins of both sexes, 8 to 15 years of age, and Kovař's (1974) study of male Czechoslovakian twins 11 to 25 years of age are considered subsequently. Several other studies from Eastern Europe are summarized in Kovař (1981a).

Variances and heritability estimates for tests of static muscular strength in male and female Polish twins are shown in Table 4. Variances were greater in DZ twins, especially males; heritabilities were moderately high, and with one exception heritability estimates tended to be higher in males than in females. Intraclass correlations and heritability estimates in Czechoslovakian male twins are summarized in Table 5. A significant role for genotype in muscular strength was clearly evident, but estimates varied with the task. Heritabilities were highest for elbow flexion ergometry, dynamometric arm strength, and the bent arm hang, and lowest for push-ups. The broad age range of the sample should be noted, and age differences between twin pairs were not considered. In con-

Table 4. Variances and heritabilities for dynamometric strength in Polish twins

Test	Males			Females		
	V_{MZ} (n = 23)	V_{DZ} (n = 19)	h^2	V_{MZ} (n = 22)	V_{DZ} (n = 36)	h^2
Elbow flexion	2.67	6.89	0.61	0.89	3.11	0.71
Elbow extension	1.61	4.45	0.64	1.02	1.90	0.46
Knee flexion	1.96	6.42	0.69	1.64	2.93	0.44
Knee extension	1.76	9.92	0.82	2.32	7.33	0.68
Back lift	19.74	99.82	0.80	9.75	35.90	0.73

Note. Data compiled from Skład (1973).

Table 5. Intraclass correlations and heritability estimates for strength in Czechoslovakian male twins 11 to 25 years of age

Test	r_{MZ}	r_{DZ}	h^2
Dynamic			
Push-ups	.65	.52	.22
Sit-ups	.85	.52	.45
Static			
Bent arm hang	.77	.64	.62
Grip	.90	.72	.45
Arm	.86	.56	.75
Trunk extension	.93	.77	.57
Elbow flexion ergometry	.87	.63	.83

Note. Data compiled from Kovař (1974); n MZ = 17, DZ = 13.

trast, Weiss (1977) reported a heritability of 0.85 for pull-ups in 10-year-old East German twins. However, this estimate was derived from correlations between 327 same-sex twin pairs and 153 opposite-sex pairs, and zygosity was not tested.

Heritability estimates from four other Eastern European studies were 0.24 and 0.71 for grip strength, and 0.11 and 0.71 for trunk extension strength. In addition, heritability estimates for strength expressed per unit body weight and for strength summed over several muscle groups tended to be similar to those for specific muscle groups, 0.64 and 0.37, respectively (Kovař, 1981a).

Among 48 pairs of male twins 9 to 11 and 15 to 17 years of age (12 MZ and 12 DZ pairs in each age group), Venerando and Milani-Comparetti (1970) reported heritabilities of 0.32 and 0.45 for right and left grip strength, and 0.49 for back extension strength. These estimates were similar to those of Kovař (Table 5), but lower than those of Skład (Table 4) for back strength.

Komi, Klissouras, and Karvinen (1973) used several measures of muscular force in 15 pairs of male and 14 pairs of female twins 10 to 14 years of age (Table 6). Although the differences in muscular forces within eight male MZ pairs were less than those within seven male MZ pairs, the intrapair variances were not statistically significant. Among female twins, however, the seven pairs of MZ twins showed as much variability in muscular forces as the seven pairs of DZ twins. Sample sizes were small, so the genetic and/or environmental significance of these sex differences is not clear.

Using a composite strength score per unit height (based on the hand grip, knee stretch, and arm bent tests), Engström and Fischbein (1977) noted more similar performances in 39 male MZ pairs than in 55 DZ pairs 18 to 19 years of age. The intrapair correlations were 0.83 and 0.47 for MZ and DZ twins, respectively. When controlling for the amount of leisure time physical activity in a subset of the twins, however, the intraclass correlation for MZ twins changed only slightly from 0.83 to 0.80, while that for DZ twins decreased from 0.47 to 0.33 (values estimated from a graph). Thus, as in studies of aerobic capacity (Bouchard & Malina, 1983b), there is a need to control for amount of habitual activity in attempting to quantify the genotypic contribution to variance in strength. Regular training has a major influence on the metabolic and functional capacity of muscle tissue.

Table 6. Within-pair differences in measures of muscular force in twins 10 to 14 years of age

Measures of Maximal force (kg)	Males				Females			
	MZ (8)		DZ (7)		MZ (7)		DZ (7)	
	M	SD	M	SD	M	SD	M	SD
Quadriceps isometric	4.2	2.2	4.9	3.9	2.3	2.1	3.9	1.9
Forearm flexor isometric	1.0	1.3	2.8	2.5	1.3	1.2	1.0	0.7
Forearm flexor eccentric	1.4	1.2	2.6	0.8	3.0	7.1	1.7	1.3
Forearm flexor concentric	2.1	2.8	3.1	3.0	1.2	0.8	1.1	0.9

Note. Data compiled from Komi et al. (1973). Number in parentheses is the number of twin pairs. None of the intrapair variances between MZ and DZ twins differed significantly.

Sibling Studies

Data on the muscular strength of siblings are not as extensive as those for twins. Montoye, Metzner, and Keller (1975) compared the aggregation of siblings for several measurements of strength: arm strength using both arms simultaneously, grip strength (right plus left), and a relative strength index based on the two measurements corrected for body size and fatness. Siblings (314 to 416 pairs) were classified into tertiles by age and sex, and similarities were then analyzed, that is, whether or not both siblings were in the same tertile. The results indicated a significant degree of similarity in strength among siblings.

Malina and Mueller (1981) considered sibling similarities in four strength tests among 114 pairs of black and 101 pairs of white siblings 6 through 12 years of age. There were no differences in sibling correlations by race, but brothers tended to resemble one another more than sisters (Table 7). The sex difference also persisted after correcting the correlations for test reliability. At face value, the heritabilities of the four static strength tests varied between 0.44 and 0.58 (Table 8). Correcting the heritabilities for body weight differences between siblings reduced the estimates, but correcting for reliability of the measurements raised the estimated heritabilities close to face value. The true heritabilities probably lie somewhere between the weight corrected and weight and reliability corrected estimates in Table 8.

More recently, Szopa (1982, 1983) reported sibling similarities in arm and grip strength in a sample of 728 urban Polish males and females 3 to 42 years

Table 7. Sibling correlations in strength of children 6 to 12 years

Test	Sis-sis (n = 49-50)	Bro-bro (n = 52-53)	Bro-sis (n = 109-112)	Total (n = 211-215)
Right grip	.19	.46*	.24*	.29*
Left grip	.21	.31*	.16	.21*
Push	.25	.45*	.13	.25*
Pull	.28*	.43*	.13	.25*

Note. Data compiled from Malina and Mueller (1981).
*p < .05.

Table 8. Estimated heritabilities of strength in children 6 to 12 years of age based on sibling correlations

Test	Face value h^2 (n = 211)	Weight corrected h^2 (n = 198)	Weight and test reliability corrected h^2 (n = 198)
Right grip	.58	.34	.55
Left grip	.44	.25	.45
Push	.50	.24	.41
Pull	.50	.17	.34

Note. Data compiled from Malina and Mueller (1981).

of age (Table 9). All but two correlations were significant, and those for arm strength were the highest. However, sex-specific sibling comparisons were not reported. Correlations for the youngest subsample (3 to 10 years) were similar to those for American black and white children 6 through 12 years of age (Table 7).

The preceding samples of siblings were rather well nourished; one can inquire about sibling similarities in nutritionally stressed populations, which include about 60 to 70% of the world's preschool and school age children. Malina and Little (1985) considered this issue in rural Indian children 6 to 14 years of age in southern Mexico. The children were living under conditions of chronic mild-to-moderate malnutrition, according to Mexican dietary criteria, while their statures and weights approximated the 5th centile of reference data for well nourished children. However, weight for height relationships were normal; that is, they were proportionately small (see Malina & Buschang, 1985).

Sibling similarities in grip strength among the Indian children are shown in Table 10. The correlations were only slightly lower than those for better nourished children (Tables 7 and 9), and the sex difference apparent in the well nourished children was not evident in the undernourished children, even after correcting for test reliability. Since strength is related to body size, the

Table 9. Sibling correlations in strength of an urban Polish sample

Strength task	Age of siblings (years)		
	3 to 10 (n = 162)	11 to 17 (n = 259)	18 to 42 (n = 93)
Right grip	.24	.17	.29
Left grip	.25	.13	(.19)
Grip/weight	(.12)	.36	.19
Arm	.38	.34	.32
Arm/weight	.38	.32	.40
Vertical jump	.35	.34	.69

Note. Data compiled from Szopa (1982); n = number of individuals in different age groups; all correlations significant, p < .05, except those in parentheses.

Table 10. Sibling correlations in grip strength of rural Zapotec children 6 to 13 years of age from the Valley of Oaxaca, Mexico

	Sis-sis (n = 43-44)	Bro-bro (n = 42-43)	Bro-sis (n = 108-109)	Total (n = 195)
Controlling for age difference between siblings				
Right grip	.28*	.26*	.05	.17*
Left grip	.14	.31*	−.05	.13*
Controlling for age difference and body weight of siblings				
Right grip	.38*	−.00	.02	.14*
Left grip	.26	.15	−.05	.11

Note. Data compiled from Malina and Little (1985).
*p < .05.

effect of controlling for body weight on sibling correlations was also considered (lower part of Table 10). Brother-brother correlations were reduced considerably, but sister-sister correlations were essentially unchanged. Body weight is a proxy for current nutritional status, and the reduced correlations between brothers may reflect the greater susceptibility of males to environmental stresses.

Parent-Child Studies

Data on parent-offspring similarities in strength are somewhat more extensive than those for siblings. Montoye et al. (1975) reported significant parent-child resemblances (475 pairs) in arm strength, grip strength, and relative strength. There did not appear to be an age effect; that is, resemblances between parents and older children (16 to 39 years) were similar to those between parents and younger children (10 to 15 years). However, female offspring appeared to resemble their parents more than did male children.

Wolański and Kasprzak (1979) compared several strength tests in parents and their offspring, 7 to 39 years, from 570 three-generation families in rural Poland. Numbers in specific age groups were small; hence, only correlations for the sample combined across age are shown in Table 11. Ages of offspring and of parents apparently were not controlled in the analysis. Nevertheless, grip strength showed a reasonable degree of parent-offspring similarity across the various pairings, and father-offspring correlations tended to be slightly higher than mother-offspring correlations. In contrast, parent-child correlations for back and shoulder strength were uniformly low. Wolański and Kasprzak (1979) suggested that parent-offspring correlations decreased during puberty, but the numbers in single age groups (generally 4 to 9) are too small to warrant such a conclusion.

In an urban Polish sample, Szopa (1982, 1983) considered age variation in parent-child relationships in strength in 1,420 individuals from 347 primarily one-generation families. Children ranged in age from 3 to 42 years, while parents ranged from 22 to 78 years. Results are summarized in Table 12. Correlations were generally low for grip strength and higher for arm strength. Parent-son correlations were higher than parent-daughter correlations for grip strength, while father-son and mother-son correlations were similar. In arm strength, mother-daughter correlations were generally similar to father-son correlations, while mother-son correlations were consistently higher than father-daughter correlations. Hence, male offspring were more similar to both parents in arm strength (to fathers more so than mothers), while females were only similar to their mothers.

Szopa's (1982, 1983) parent-offspring correlations for grip strength in an urban Polish sample were consistently lower than those reported by Wolański and Kasprzak (1979) for a rural Polish sample (Table 11). However, the correlations for shoulder strength in the latter were less than those for arm strength in the former. The difference may reflect variation in socioeconomic status and life-style (including physical activity, which was probably greater in the rural sample), as well as sample size and test procedures. On the other hand, the pattern of correlations for arm and shoulder strength was similar in both studies. Girls were more similar to their mothers and sons were more similar

Table 11. Parent-offspring correlations for strength in a rural Polish sample

Test	F-S (n = 59-63)	M-S (n = 76-82)	MP-S (n = 43-44)	F-D (n = 49-53)	M-D (n = 62-71)	MP-D (n = 30-32)
Right grip	.55**	.46**	.46**	.62**	.46**	.57**
Left grip	.50**	.48**	.43**	.43**	.37*	.38*
Shoulder pull	.20	.14	.26	−.07	.27*	−.00
Back lift	−.19	−.13	−.24	.02	−.03	−.19

Note. Data compiled from Wolański and Kasprzak (1979); age of offspring, 7.5 to 39.4 years.
*p < .05; **p < .01.

Table 12. Parent-offspring correlations for strength in an urban Polish sample

	F-S			M-S			MP-S		
	A	B	C	A	B	C	A	B	C
Right grip	.23	(.11)	.29	.20	(.13)	.17	.26	(.12)	.30
Left grip	.20	.19	.20	.21	.21	.20	.25	.26	.24
Grip/weight	(.15)	.21	.23	(.18)	(.16)	.22	.22	(.17)	.32
Arm	(.19)	.35	.33	.24	.21	.20	(.19)	.38	.28
Arm/weight	(.11)	.30	.24	.34	.24	(.12)	(.14)	.27	.17
Vertical jump	(.17)	.48	.46	.38	.44	.33	.36	.54	.54
	F-D			M-D			MP-D		
	A	B	C	A	B	C	A	B	C
Right grip	(.13)	(.07)	(.03)	(.10)	(.03)	.30	(.16)	(.06)	(.09)
Left grip	(.03)	(−.04)	(.00)	(.07)	(.03)	(.19)	(.01)	(−.04)	(−.03)
Grip/weight	(−.06)	.17	(.02)	(−.10)	(.11)	(.02)	(−.06)	.18	(−.03)
Arm	(.14)	(.01)	(.14)	.31	.31	.42	(.19)	.21	.33
Arm/weight	(.09)	(.05)	(.09)	.28	.33	.50	.25	.24	.42
Vertical jump	.24	.32	.37	.21	.31	.48	.21	.39	.47

Note. Data compiled from Szopa (1982); all correlations significant, $p < .05$, except those in parentheses; age groups: A = 3-10, B = 11-17, C = 18-42 years.

to their fathers; but father-daughter correlations were lower than mother-son correlations in both studies.

Szopa (1982, 1983) also reported heritabilities for the strength tests (Table 13). The estimates included a correction for spouse similarity, that is, assortative mating. Heritabilities tended to be higher for males than females for grip strength but were similar for arm strength. The sex difference in heritabilities for grip strength is of interest in that it suggests little if any genetic influence in females, contrary to the general notion that females are genetically more buffered from environmental stresses. Although there was variation among estimated heritabilities with age of the sample, the estimates tended to be higher in the older groups. The heritabilities for grip strength in children 3 to 10 years of age were lower than those for children 6 to 12 years of age studied by Malina and Mueller (1981), while the estimates for arm strength were similar to those for shoulder pulling and pushing strength (Table 8). Note, however, that the heritabilities in the latter study were derived from sibling correlations corrected for body weight differences between siblings and test reliability.

Kovař (1981b) considered similarities between parents and their 16- and 17-year-old sons in grip and back strength. The father-son and midparent-son correlations were significant for grip strength (0.31 and 0.33, respectively), while the mother-son correlation was not (0.07). The former correlations were higher than those reported by Szopa (1982, 1983) for 11- to 17-year-old children and their parents (Table 12), while the mother-son correlation was lower. All three correlations for grip strength were lower than those reported by Wolański and Kasprzak (1979) (Table 11). For back strength, on the other hand, Kovař (1981b) found low parent-son similarities (father-son, 0.07; mother-son, 0.16; midparent-son, 0.07). These correlations differed from those reported by

Table 13. Estimated heritabilities for strength in an urban Polish sample

Age groups	Right grip	Left grip	Grip/ weight	Arm	Arm/ weight	Vertical jump
Males						
3-10	.37	.33	.27	.37	.38	.44
11-17	.21	.32	.30	.48	.46	.68
18-42	.40	.32	.37	.45	.31	.59
Females						
3-10	.20	.00	.00	.38	.32	.33
11-17	.09	.00	.23	.27	.32	.47
18-42	.29	.15	.03	.48	.50	.63

Note. Data compiled from Szopa (1982); heritabilities include an adjustment for assortative mating/spouse similarity.

Wolański and Kasprzak (1979), who reported negative correlations between parents and offspring in back strength (Table 11). Variation in correlations among these studies probably reflects age variation in the samples of parents and their children. In another study of a small sample (*N* = 15) of 16- and 17-year-old boys and their fathers, Kovař (1981a) reported a correlation of 0.68 for elbow flexion strength per unit body weight. This correlation was more than twice the value reported by Szopa (1982, 1983) for arm strength per unit weight in 11- to 17-year-old sons and their fathers (Table 12).

Runs, Jumps, and Throws

Tests of running, jumping, and throwing are perhaps the most commonly used tasks in assessing gross motor performance. They are generally reliable and do not require elaborate equipment. Such tests are sometimes referred to as indicators of explosive strength, that is, the ability of muscles to release maximal force in the shortest possible time (Clarke, 1967). Thus, the runs or dashes require the body to move as rapidly as possible over a short distance.[2] Jumps usually require the body to move either vertically or horizontally through space from a standing position. Throws require an object to be projected forcefully, usually as far as possible. Of course there are variants of these basic tasks, each with its own requirements, such as the shuttle run, the running long jump, and the throw for accuracy.

Twin Studies

In an early study of twins 4.0 to 4.5 years of age, Mirenva (1935) noted smaller within-pair differences among four pairs of MZ twins compared to six pairs of DZ twins in jumping and throwing tasks. Mean within-pair differences for MZ and DZ twins, respectively, were 1.7 and 7.6 cm in height of a jump, 0.25 and 1.47 in the score on a target throw, and 15.0 and 23.8 cm in the deviation from a target on a ball rolling task.

Results of three Japanese studies of twins which utilized tests of running, jumping, and throwing are summarized in Table 14. The samples and methods

---

[2]Runs for longer distances, for example, 600 yards, mile, and so on are ordinarily classified as endurance tasks and will not be considered.

Table 14. Tests of running, jumping, and throwing in Japanese twins: Mean percent deviation/difference within pairs

Motor tests	Ishidoya (1957) Like-sex		Kimura (1956)				Mizuno (1956)	
			Males		Females		Male	Unrelated
	MZ	DZ	MZ	DZ	MZ	DZ	MZ	Pairs
	(57)	(44)	(34-143)	(6-55)	(35-137)	(17-79)	(15)	(27-29)
50-meter dash	2.92	5.01	2.3	3.2	1.8	4.5	1.83	8.50
Long jump	2.87	4.95	3.4	3.4	3.2	5.7	8.62	9.70
Vertical jump			4.8	6.7	3.9	6.8	11.94	16.28
Distance throw	4.79	8.81					10.62	22.99

Note. Numbers in parentheses are the number of twin pairs.

of analysis have been described earlier. Kimura (1956) reported generally smaller percentage deviations within MZ twin pairs on the vertical jump, standing long jump, and 50-meter dash than within DZ pairs of both sexes. However, the differences were statistically significant only for female MZ and DZ twins in the vertical jump. Using similar methods, Ishidoya (1957) noted significantly smaller percentage deviations within MZ pairs on the standing long jump, 50-meter dash, and ball throw for distance. The estimated genetic contribution to performance on the three tasks was about 40 to 45%. In Mizuno's (1956) analysis, male MZ twins showed smaller within-pair mean percentage differences than pairs of grade-matched unrelated boys on the vertical and standing long jumps, the baseball throw for distance, and the 50-meter dash. Differences were less for the two jumps than for the run and throw. Partial correlations, controlling for age, did not differ between MZ males and females for the jump and the throw, but were higher in MZ males for the 50-meter dash. Partial correlations for grade-matched pairs of unrelated boys were uniformly low.

Estimated heritabilities for runs, jumps, and throws in three samples of Eastern European twins are summarized in Table 15. The samples and methods have been described earlier. Although the heritabilities were estimated by different methods, the data suggest a greater genetic influence in the runs and jumps than in the throws. The triple jump is an exception, but it is not entirely an explosive jump. It requires coordinating a hop, step, and then a jump, and is probably more amenable to training influences. The sex-specific analysis of two tasks in Polish twins indicated no clear pattern of sex differences in heritabilities.

In a summary of heritability estimates from other Eastern European studies of twins between 5 and 17 years of age, Kovař (1981a) reported heritabilities of 0.45, 0.59, and 0.74 for the long jump, 0.82 for the vertical jump, and 0.14 for the shotput. Estimated heritabilities for the dash varied with distance, that is, 0.83 for a 20-meter dash, 0.62 for a 30-meter dash, and 0.45 and 0.91 for a 60-meter dash. These estimates are generally comparable to those in Table 15. However, the estimated heritabilities for dashes of different distances do not support the suggestion of Wolański, Tomonari, and Siniarska (1980) that heritability is highest for shorter dashes and decreases with the longer dashes. In contrast to the above, Komi et al. (1973) reported an almost perfect heritabili-

Table 15. Estimated heritabilities for tests of running, jumping, and throwing in twins

Author/test	Males			Females		
Skład (1973)[a]	V_{MZ}	V_{DZ}	h^1	V_{MZ}	V_{DZ}	h^2
60-meter dash	0.67	2.31	.72	0.63	3.23	.80
vertical jump	6.43	25.26	.75	7.84	21.39	.63
Kovař (1974)[b]	r_{MZ}	r_{DZ}	h^2			
shuttle run	.93	.52	.90			
vertical jump	.90	.42	.86			
medicine ball throw	.92	.61	.60			
Weiss (1977)[c]			r_{SS}	r_{OS}	h^2	
60-meter dash			.69	.38	.85	
long jump			.60	.34	.74	
triple jump			.60	.21	.66	
cricket ball throw			.54	.11	.54	
shotput			.57	.33	.71	

[a]Ns per task: Male MZ 15, 23; DZ 12, 19; female MZ 12, 22; DZ 14, 36, respectively.
[b]Ns: MZ 17, DZ13; intraclass correlations.
[c]SS = same sex, n 327 pairs; OS = opposite sex, n 153 pairs.

ty (0.992) for maximum muscular power derived from a test of maximal running velocity on a staircase in eight MZ and seven DZ pairs of male twins. However, seven pairs of female MZ twins were not any more alike in this test than seven pairs of female DZ twins.

Skład (1972) also considered the genotypic contribution to the biomechanical composition of running performance in 27 pairs of MZ and 23 pairs of DZ twins 11 to 15 years of age. The data were based on cinematographic analysis of certain aspects of running style during a 60-meter dash. Variance and estimated heritabilities for stride length and tempo of the run are shown in Table 16, while the same statistics for limb and trunk angles at different phases of the 60-meter dash are shown in Table 17. Clearly, the kinematic structure of the 60-meter dash was more similar in MZ than in DZ twins. Variances within DZ pairs were greater than within MZ pairs, especially in males. Variances were only slightly greater in female DZ than in female MZ twins. Hence, the heritability estimates for stride length, tempo (speed and number of strides over 60 meters), and limb and trunk angles at different phases of the dash were lower in females than in males. It would thus appear that the genotypic contribution to variation in the kinematic structure of the dash is greater in boys than in girls. This suggests that the running performance of girls is more amenable to environmental influences, including, of course, social and motivational factors for an all-out performance.

Sibling and Parent-Offspring Studies
Sibling similarities among children 6 through 12 years of age in tests of running, jumping, and throwing are summarized in Table 18. Brother pairs were more alike than sister pairs in the 35-yard dash and the softball throw for distance, with little difference in the standing long jump. The sex difference in sibling correlations also persisted after correction for test reliability (Malina

& Mueller, 1981). At face value, heritability estimates for the three motor tasks were low for the dash (0.26) and jump (0.22), and moderate for the throw (0.56). However, adjusting the estimates for body weight differences between siblings and for test reliability increased the estimates considerably (Table 19), to levels comparable to those based on twin studies (Table 15).

Table 16. Variances and heritabilities for certain aspects of a 60-meter dash

	Males			Females		
	V_{MZ} (n = 15)	V_{DZ} (n = 11)	h^2	V_{MZ} (n = 12)	V_{DZ} (n = 12)	h^2
Stride length, R leg take-off	23.12	81.76	0.72	5.02	14.75	0.66
Stride length, L leg take-off	7.58	160.11	0.95	1.77	14.29	0.88
Mean stride length	9.46	106.33	0.91	2.31	13.43	0.83
Tempo	0.68	2.67	0.74	0.50	1.33	0.63

Note. Data compiled from Skład (1972).

Table 17. Variances and heritabilities for limb and trunk angles at different phases in a 60-meter dash in Polish twins

Angle between:	Males			Females		
	V_{MZ} (*n* = 15)	V_{DZ} (*n* = 11)	h^2	V_{MZ} (*n* = 12)	V_{DZ} (*n* = 12)	h^2
Take-off right leg						
R arm & forearm	26.45	247.75	0.89	91.55	370.21	0.75
R arm & trunk	6.32	27.49	0.77	4.69	18.20	0.74
Trunk & L thigh	27.41	49.50	0.45	44.30	37.96	
L thigh & leg	46.55	323.39	0.86	43.55	145.63	0.70
R thigh & leg	10.36	37.61	0.72	34.35	23.13	
R & L thighs	25.50	136.72	0.81	37.60	47.79	0.21
Trunk & perpendicular	22.73	13.67		19.80	14.75	
Support phase						
R arm & forearm	62.27	217.28	0.71	126.90	230.21	0.45
Trunk & R thigh	57.23	217.28	0.74	90.95	136.67	0.34
R thigh & leg	88.27	160.83	0.45	162.60	103.92	
Trunk & perpendicular	13.18	19.28	0.32	9.60	35.42	0.73
Take-off left leg						
R arm & forearm	26.23	214.67	0.88	102.45	474.54	0.79
R arm & trunk	44.32	132.89	0.67	41.45	158.17	0.74
Trunk & R thigh	19.95	58.17	0.66	30.35	17.87	
R thigh & leg	20.59	123.78	0.83	102.50	176.29	0.42
L thigh & leg	10.86	56.44	0.81	22.20	36.00	0.38
R & L thighs	12.32	87.61	0.86	36.10	36.21	
Trunk & perpendicular	11.73	19.28	0.39	2.10	13.17	0.84

Note. Data compiled from Skład (1972).

Table 18. Sibling correlations in the motor performance of children 6 to 12 years of age

Test	Sis-sis (n = 49-50)	Bro-bro (n = 52-53)	Bro-sis (n = 109-112)	Total (n = 211-215)
35-yard dash	.09	.28*	.08	.13
Long jump	.20	.16	.02	.11
Distance throw	.19	.36*	.28*	.28*

Note. Data compiled from Malina and Mueller (1981).
*$p < 0.05$.

Table 19. Estimated heritabilities of motor performance in children 6 to 12 years of age based on sibling correlations

Test	Face value h^2 (n = 211)	Weight corrected h^2 (n = 198)	Weight and test reliability corrected h^2 (n = 198)
35-yard dash	.26	.25	.81
Long jump	.22	.22	.60
Distance throw	.56	.48	.67

Note. Data compiled from Malina and Mueller (1981).

Szopa (1982, 1983) reported significant correlations of 0.35, 0.34, and 0.69 for the vertical jump in three age groups of urban Polish siblings, 3 to 10 years, 11 to 17 years, and 18 to 42 years, respectively (see Table 9). The correlation in the youngest age group (0.35) was about three times that reported for the standing long jump in Philadelphia children 6 through 12 years of age (0.11) (Table 18). Sibling similarities in the 35-yard dash, standing long jump, and ball throw for distance among a sample of mild-to-moderately undernourished children 6 through 13 years of age are shown in Table 20. With one exception, the correlations were generally lower than those for well nourished children on the same three tasks (Table 18). Interestingly, sisters resembled each other more than brothers in the undernourished sample. Since performance is related to body size, the effect of controlling for body weight, a proxy for nutritional status, was also considered (lower part of Table 20). However, the correction changed the sibling correlations only slightly.

Data on parent-child correlations for runs, jumps, and throws are not extensive. Many of the tests used with children and youth are simply not suitable for adults. Cratty (1960) compared the performances of 24 college-age males with that of their fathers who had attended the same college 34 years earlier (i.e., when the fathers were of college age). Father-son correlations were 0.86 for the running long jump, 0.59 for the 100-yard dash, and 0.04 for the fence vault. Hence, fathers and sons attained similar performances in two of the tasks. Body size was not controlled in the analysis, and this may be significant since secular changes may have occurred between 1925 and 1959. Nevertheless,

Table 20. Sibling correlations in motor performance of rural Zapotec children 6 to 13 years of age from the Valley of Oaxaca, Mexico

	Sis-sis (n = 36-43)	Bro-bro (n = 39-42)	Bro-sis (n = 101-106)	Total (n = 179-191)
Controlling for age difference between siblings				
35-yard dash	.48*	.11	.08	.17*
Long jump	.09	.02	.02	.03
Distance throw	.18	.06	-.22*	-.05*
Controlling for age difference and body weight of siblings				
35-yard dash	.51*	.11	.13	.20*
Long jump	.14	.04	.03	.01
Distance throw	.21	.03	-.13	.03

Note. Data compiled from Malina and Little (1985).
*p < 0.05.

it is of interest that the fathers performed better than their sons in the 100-yard dash and fence vault, while they did not differ in the running long jump.

Szopa (1982, 1983) also included parent-offspring similarities in the vertical jump; results have already been reported in Table 12. All correlations except one were significant, and midparent-child correlations tended to be higher than father-child or mother-child correlations. However, in the oldest group of siblings (18 to 42 years), sons were more similar to their fathers while daughters were more similar to their mothers. Estimated heritabilities for the vertical jump (see Table 13) tended to be higher in young males (3 to 10 years and 11 to 17 years) than in young females, while those in older individuals (18 to 42 years) differed only slightly between the sexes. These heritabilities were only slightly lower than those reported for various jumping tests in twins (Table 15) and other siblings (Table 19). However, Szopa's estimates included an adjustment for spouse similarity and are thus more conservative since assortative mating in the Polish sample tended to increase the heritability estimates.

Balance

Balance is a skill that requires a combination of gross and fine motor control in maintaining equilibrium. Hence, balance tests are often included in surveys of motor performance as balance is an essential component of skillful performance in a variety of fundamental and specialized skills. Heritability estimates for several balance tasks in twins are summarized in Table 21. Estimates ranged from 0.27 to 0.86 and varied with the test used. Skład's (1973) data also suggested a sex difference; heritabilities were higher for males than females in a rail balance task.

Wolański and Kasprzak (1979) considered parent-offspring similarities in three balance tasks in a rural Polish sample (Table 22). Ages of offspring varied between 6 and 19 years but ages of parents were not reported. Correlations were low for beam walking and the timed turning balance test (number of turns in 10 seconds). For the former, father-son correlations were highest while father-daughter and mother-daughter correlations were similar. For the latter,

Table 21. Estimated heritabilities for various balance tests in twins

Author/test	Sex	n_{MZ}	n_{DZ}	h^2
Vandenberg (1962)				
Seashore beam balance	M + F	41	32	.48
Skład (1973)				
one foot lengthwise on rail, eyes open				
Right foot	M	23	19	.86
	F	22	36	.63
Left foot	M	23	19	.81
	F	22	36	.76
Williams & Heartfield (1973)				
Bachman ladder climb	M + F	15	8	.46
Williams and Gross (1980)				
stabilometer	M + F	22	41	.27

Table 22. Parent-offspring correlations for balance in a rural Polish sample

	F-S (n = 93-96)	M-S (n = 110-125)	MP-S (n = 61-65)	F-D (n = 90-94)	M-D (n = 106-118)	MP-D (n = 62-63)
Beam walking	.21*	−.01	.08	.13	.12	.17
Turning, no./sec	.10	.16	.17	.23*	.14	.14
Turning, no. turns	.41**	.28**	.37**	.12	.31**	.34**

Note. Data compiled from Wolański and Kasprzak (1979); age of offspring, 6.5 to 19.5 years.
$*p < .05$; $**p < .01$.

on the other hand, father-daughter correlations were the highest and mother-son and mother-daughter correlations were similar. In contrast, the untimed turning balance test (number of full circles) showed the highest parent-offspring correlations. The father-son correlation was the highest, while mother-son, mother-daughter, and midparent-child correlations were similar. Thus, with time or speed as a factor there was little parent-offspring resemblance in balance while turning, while with the time factor removed, parent-offspring similarities were significant. Ages of offspring and parents were not statistically controlled in the analysis. Nevertheless, parent-offspring correlations tended to be somewhat lower when offspring were between 11 and 14 years of age (i.e., circumpuberally), although numbers in some age cells were small.

Speed of Limb Movement, Manual Dexterity, and Coordination

The movement tasks considered in this section are often labeled as fine motor, that is, they require precision of movement, sometimes with speed. The evidence is primarily from twin studies and indicates a significant genetic effect in a variety of tasks such as pursuit rotor tracking, tapping speed, hand steadiness, maze tracing, and so on. Results are summarized in Tables 23 and 24. Heritabilities varied with the task and indicated generally higher genetic contributions for performances with the right hand than the left (Table 23).

Table 23. Heritability estimates and F-ratios for motor skills tests in same-sex twins

Test	Right hand h^2	F	Left hand h^2	F
Mirror drawing	.70	3.38**	.24	1.31
Tweezer dexterity	.71	3.40**	.63	2.73**
Santa Ana dexterity	.58	2.41**	.05	1.05
Hand steadiness	.37	1.59	.17	1.21
Rotary pursuit	.52	2.08*	.32	1.46
Card sorting	.61	2.57**	.71	3.42**

Note. Data compiled from Vandenberg (1962); n DZ = 30-35, n MZ = 40-45, high school age.
*$p < .05$; **$p < .01$.

Table 24. Variances and heritabilities for limb speed, accuracy, and dexterity in Polish twins

Test	Males V_{MZ}	V_{DZ}	h^2	Females V_{MZ}	V_{DZ}	h^2
	$(n = 23)$	$(n = 19)$		$(n = 22)$	$(n = 36)$	
Upper limb plate tapping						
Right	17.70	130.53	.86	17.84	93.11	.81
Left	28.96	202.95	.86	33.02	86.22	.62
One foot tapping						
Right	2.76	27.74	.90	9.55	26.93	.65
Left	4.74	23.34	.79	7.52	27.08	.72
	$(n = 15)$	$(n = 12)$		$(n = 11)$	$(n = 13)$	
Arm movement speed and accuracy	45.27	52.00	.13	36.68	85.89	.57
Manipulative dexterity	8.50	17.17	.51	12.68	20.39	.38

Note. Data compiled from Skład (1973).

Speed of limb movement, however, did not appear to show right-left differences in heritability (Table 24). In Skład's (1973) sample of Polish twins, estimated heritabilities tended to be higher in male twins than in female twins for speed and manipulative tasks, while the opposite was true for the task requiring speed and accuracy of arm movement (Table 24).

In other twin studies, Kovař (1974) reported moderate heritabilities of 0.48 and 0.64 for a manual coordination task in male twins 11 to 25 years of age. These estimates are well within the range of those indicated in Tables 23 and 24. Kimura (1956) noted smaller within-pair percentage deviations in MZ than in DZ twins of school age in tapping speed, manual dexterity, and finger dexterity. The differences between MZ and DZ twins were not significant, however, except for manual dexterity in males. In addition, the data for three age groups suggested age-associated variation in the three tasks, more so for males than females.

Parent-offspring correlations for fine motor tasks in a rural Polish sample are shown in Table 25. Eye-hand coordination in a tracing task appeared to

Table 25. Parent-offspring correlations for arm movements in a rural Polish sample

Test	F-S	M-S	MP-S	F-D	M-D	MP-D
	(n = 123)	(n = 160-162)	(n = 95)	(n = 107)	(n = 138-140)	(n = 81)
Eye-hand coordination (tracing task)						
Right, no. errors	.08	.21**	.31**	.30**	.24**	.39**
Left, no. errors	.20*	.38**	.39**	.19	.24**	.31**
Right, errors/sec	.37**	.14*	.38**	.29**	.36**	.45**
Left, errors/sec	.27**	.08	.29**	.23*	.34**	.44**
	(n = 116)	(n = 163)	(n = 87)	(n = 99)	(n = 144)	(n = 72)
Movement accuracy (arm lift to 120°)						
Right	.00	-.02	.08	.02	.12	.22
Left	-.03	-.08	-.10	.08	.13	.18

Note. Data compiled from Wolański and Kasprzak (1979); age of offspring 6.5 to 39.4 years.
$*p < .05$; $**p < .01$.

Table 26. Parent-son correlations in fine motor tasks

Test	F-S	M-S	MP-S
Tapping speed	.12	.26*	.07
Maze, time	.24	.22	.34*
Maze, errors	.10	.24	.14

Note. Data compiled from Kovař (1981); n = 60 sons 16 to 17 years.
*$p < 0.05$.

show a familial pattern, but parent-offspring correlations varied by pairing. Mother-daughter correlations for errors per unit time were greater than mother-son correlations, while father-son correlations for errors per unit time were only slightly greater than father-daughter correlations. For the number of errors, only mother-child correlations were consistently significant. There was no consistent tendency toward higher correlations for the right or left hand in number of errors on the tracing task. However, correlations for errors per unit time tended to be higher for the right hand, more so for males than females. Movement accuracy, a task that reflects proprioceptive feeling in lifting the arm to 120 degrees, showed no parent-offspring similarities. The same was true for arm lifting to 60 degrees (not shown in the table).

Kovař (1981b) considered performance similarities in three fine motor tasks in 16- and 17-year-old boys and their parents (Table 26). Correlations were generally low, and mother-son correlations were higher in two of the three tasks. The correlations were generally lower than those reported by Wolański and Kasprzak (1979) for a tracing task (Table 25). The difference most likely reflects age variation in the samples compared.

Miscellaneous Motor Tasks

The number of possible motor tasks available for testing is almost limitless. Some tests include a series of stunts, and a performance score for each of the stunts yields a single composite score. Sklad (1973), for example, reported moderately high heritabilities of 0.80 and 0.74 for male and female twins, respectively, on the Johnson test, which supposedly measures inherent capacity for neuromuscular skill (Larson & Yocom, 1951). The test consists of 10 stunts (e.g., stagger skips, half turns right and left, and so on), and each is scored on a 10-point basis. Vickers, Poyntz, and Baum (1942) used a modified Brace test, which also consists of a series of graduated stunts scored on a pass-fail basis. The authors noted clear similarities in the performances of seven pairs of twins and siblings on this test compared to unrelated pairs of children 5 to 9 years of age. Such stunt tests as the Johnson and Brace tests place a premium on balance, agility, and coordination. Mizuno (1956) included the Burpee test (squat thrusts) in his study of Japanese twins. The mean percentage difference in unrelated pairs of boys was more than three times that observed in pairs of male MZ twins. Using a hopping test, Ishidoya (1957) reported a mean percentage deviation in like-sex DZ twins which was significantly greater than that in like-sex MZ twins. On the other hand, Ishidoya noted no differences between MZ and DZ twins in ball striking and rope jumping tasks.

Assortative Mating for Motor Skills

Studies of biological relatives should allow for assortative mating, the tendency of individuals to select mates with similar characteristics. Positive assortative mating for ability, education, religion, and so on is reasonably well established (Garrison, Anderson, & Reed, 1968; Harrison, Gibson, & Hiorns, 1976; Johnson, DeFries, Wilson, McClearn, Vandenberg, Ashton, Mi, & Rashad, 1976), as is that for physical characteristics (Roberts, 1977; Spuhler, 1968; Susanne, 1967). Husband-wife correlations for body size and related measurements tend to be low to moderate in populations of primarily European ancestry.

Spouse similarities are also evident in motor performance variables. Among American couples, Montoye et al. (1975) observed significant husband-wife similarities in grip, arm, and relative strength. Correlations were not reported. Rather, spouses were classified into tertiles, and similarities were then analyzed, that is, whether both spouses were in the same tertiles. Younger couples (16 to 39 years) showed greater similarity in strength than older couples (40+ years).

Spouse correlations for motor characteristics in three Eastern European samples are summarized in Table 27. Szopa's (1982, 1983) sample of urban Poles included 347 families, and this number presumably approximates the

Table 27. Husband-wife correlations in motor tasks

Wolański (1973)[a]		Szopa (1982)[b]		Kovař (1981)[c]	
Strength		Strength		Strength	
R grip	.74	R grip	.15	Grip	.26
L grip	.77	L grip	.26	Back	.26
Shoulder	.53	Grip/weight	.23	Tapping speed	.25
Back	.44	Arm	.17	Maze, time	.44
Reaction time		Arm/weight	.17	Maze, errors	(.07)
Optical	.64	Vertical jump	.35		
Acoustic	.52				
Tactile	.57				
Balance					
Beam walk	(−.50)				
Turn, time	(.25)				
Turn, number	(.04)				
Eye-hand coordination					
Errors, R	.55				
L	.53				
Time, R	.35				
L	.24				
Movement accuracy					
R	(.16)				
L	(.14)				

Note. All correlations $p < .05$, except those in parentheses.
[a]Ns vary with task, 36 to 72.
[b]Based on 347 families.
[c]60 spouse pairs.

number of spouse pairs. Wolański's (1973) rural Polish sample included 36 to 72 spouse pairs, the numbers varying with the task, while Kovař's (1981) urban Czechoslovakian sample included 60 spouse pairs. Spouse correlations for strength tended to be low to moderate, although those reported by Wolański (1973) were rather high compared to the other two studies. Malina, Selby, Buschang, Aronson, and Little (1983) reported similar correlations for grip strength (right 0.29, left 0.27) among spouses in a rural Indian community in southern Mexico, while Baldwin and Damon (1973) noted more variable spouse correlations for grip strength (0.10, 0.96, 0.47, 0.35) in four Solomon Island populations.

Data on husband-wife similarities in movement tasks are less available than those for strength. Nevertheless, most of the correlations shown in Table 27 were in the same range as those for strength. The data thus indicate positive assortative mating for a number of motor characteristics. The magnitude of the correlations in many instances are similar to those reported for physical characteristics (Roberts, 1977; Spuhler, 1968; Susanne, 1967). Such deviations from random mating in turn must be considered in evaluating the genetics of motor performance.

Associated Factors: Covariates of Motor Performance

Many of the motor tasks considered in the preceding section show a moderate to moderately high degree of heritability. However, most tests are done on a single occasion and do not consider improvement in performance over time. Improvement in performance may include a learning component and a genetic effect. The same applies to muscular strength, which is also influenced by regular training. The training response in turn may also include a genetic component (Bouchard & Malina, 1983b; Bouchard & Lortie, 1984). Perceptual and perceptual-motor factors are often implicated as significant in motor performance (see Williams, 1983), while the association between performance and size, physique, and body composition is well established (Boileau & Lohman, 1977; Malina, 1975). The genotypic contribution to variation in morphological characteristics is rather well documented (see Bouchard & Malina, 1983b; Bouchard & Lortie, 1984), and will not be considered. Only estimates of the genetic component of variance in motor learning, muscular training, and perceptual factors will be considered.

Motor Learning

Studies of motor learning in twins have a long history. Early motor development research focused in part on the effects of early practice on the development of motor skills (i.e., learning versus maturation). This focus is well illustrated in the early co-twin control studies of Gesell and Thompson (1929), Hilgard (1933), McGraw (1935), and Mirenva (1935). One twin was given specific practice or training in one or several motor activities while the other twin received no special training or practice. Although the studies showed some improvement with training, maturational processes, which were viewed as genetically based, were seemingly more important than the special training and practice. Nevertheless, qualitative differences in creeping and walking favored the trained twin in McGraw's (1935) study of Johnny and Jimmy. Mattson (1933) also suggested that early training may be more effective with complex rather than simple skills. This would seem to be indicated in the early mastery of roller skating demonstrated by McGraw's trained twin.

McNemar (1933) considered the influence of practice on three fine motor skills in 47 MZ and 48 DZ male twin pairs of junior high school age (Table 28). Practice did not change the degree of similarity between MZ twins in pursuit rotor, spool packing, and card sorting performances. However, it did increase the degree of similarity between DZ twins in the pursuit rotor and spool packing tasks, while reducing the similarities in card sorting. Due to the change in intraclass correlations with practice, heritability estimates differed on the initial and final tests. They decreased from 0.78 to 0.67 on the pursuit rotor and from 0.30 to 0.00 on spool packing, but remained the same on the initial and final tests for card sorting, 0.44 and 0.43, respectively.

In another early study of the effects of practice on fine motor skills of twins, Brody (1937) reported closer resemblances between 29 MZ than 33 DZ male twin pairs 8 to 14 years of age on a mechanical ability test, and a series of six practice trials did not influence the degree of twin similarity. Wilde (1970), however, reanalyzed Brody's (1937) data and noted a greater increase in MZ than DZ similarity over the six practice trials. Hence, the estimated heritability increased from an insignificant 0.35 on the first trial to a significant 0.59 on the sixth trial.

The reanalysis of early data by Wilde (1970) emphasizes the variability in heritability estimates based on the first trial only and the need to consider practice or learning effects. More recently, Marisi (1977) noted a relatively high heritability for initial performance on a pursuit rotor task (0.96) in 35 MZ and 35 DZ pairs of male and female twins 11 to 18 years of age (Table 29). Heritability decreased to 0.45 over 30 practice trials. After a 30-minute rest period the heritability estimate increased to 0.85, but it subsequently declined to 0.58 over 20 additional trials. Thus, with practice the strength of the estimated genetic contribution systematically diminished. These results, which indicate changing estimates of heritability in fine motor tasks with practice, may also apply to gross motor tasks and perhaps to performances of young children where reliabilities may be lower.

Table 28. Intraclass correlations and heritability estimates for the first two and last two segments of a practice series in three motor tasks in male twins

Segment	Pursuit rotor			Spool packing			Card sorting		
	r_{MZ}	r_{DZ}	h^2	r_{MZ}	r_{DZ}	h^2	r_{MZ}	r_{MZ}	h^2
1	.88	.45	.78	.56	.37	.30	.75	.56	.44
2	.86	.46	.75	.57	.54	.07	.74	.47	.51
6	.86	.61	.63	.52	.59	.00	.62	.45	.30
7	.87	.60	.67	.54	.55	.00	.71	.49	.43

Note. Data compiled from McNemar (1933); n MZ = 45, n DZ = 46, mean age 14.3 years.

Using a stabilitometer task, Williams and Gross (1980) reported different results in 22 MZ and 41 DZ pairs of male and female twins 11 to 18 years (Table 30). Heritability was low on the first day of practice (0.27) but increased to 0.69 on the second day and then remained close to this level over the next 4 days. These results would seem to suggest that low heritability when learning the stabilitometer task was relatively rapid early in practice. However, a stable heritability level was reached rather quickly. These results thus differ from those of McNemar (1933) and Marisi (1977), and emphasize perhaps the specificity of heritability in particular motor skills and of course the need to consider the learning and practice effects.

Skład (1975) compared the learning curves and rates of improvement in 24 MZ and DZ like-sex twin pairs 9 to 13 years of age during the course of learning four tasks: plate tapping with the hand, one foot tapping, mirror tracing, and ball tossing for accuracy. The tasks were performed daily over a period of 10 to 14 days under the same conditions. The performance curves were

Table 29. Intraclass correlations and heritability estimates over trial blocks in the performance of twins on a pursuit rotor apparatus

Trial blocks	r_{MZ}	r_{DZ}	h^2
1-5	.73	.25	.96
6-10	.84	.46	.77
11-15	.81	.32	.98
16-20	.80	.50	.61
21-25	.87	.57	.61
26-30	.85	.63	.45
30-minute retention interval			
31-35	.87	.45	.85
36-40	.90	.56	.67
41-45	.88	.48	.79
46-50	.85	.56	.58

Note. Data compiled from Marisi (1977); n MZ = 35, n DZ = 35 pairs, 11 to 18 years of age.

Table 30. Intraclass correlations and heritability estimates over days in the performance of twins on a stabilometer task

Day	r_{MZ}	r_{DZ}	h^2
1	.51	.33	.27
2	.73	.28	.69
3	.70	.33	.56
4	.75	.31	.66
5	.77	.33	.65
6	.78	.36	.67
1-6	.75	.34	.62

Note. Data compiled from Williams and Gross (1980); n MZ = 22, n DZ = 41 pairs, 11.5 to 18.3 years of age.

more similar in MZ than in DZ twins, and more similar in MZ females than in MZ males. The sex difference was marked in mirror tracing and ball tossing but less marked in the tapping tasks. Mathematical fitting of the learning curves resulted in three parameters for each skill: level, rate, and final stage of learning. Two of the parameters, level and rate of learning, were generally more similar in MZ twins. Intrapair correlations for these parameters are summarized in Table 31 for each motor task. For the first parameter, level of learning, intrapair correlations were higher in MZ twins and slightly higher in male MZ than in female MZ twins for the speed tapping and ball tossing tasks. Sex differences in the correlations were minor in mirror tracing. However, it should be noted, that in the time score for mirror tracing and in ball tossing with the right hand, MZ and DZ intrapair correlations were about the same.

For the second parameter, rate of learning, the intrapair correlations were generally greater in MZ than in DZ twins, with the exception of foot tapping with the right foot and ball tossing with the left hand in boys, and the time score for mirror tracing in girls. For the third parameter, final stage of learning, intrapair correlations (not shown in the table) were more variable between tasks and the sexes.

Therefore, it is apparent that estimates of the genetic contribution to the learning of motor skills varies from task to task. Sex differences also vary from task to task as suggested by Skład's (1975) study. Hence, studies combining male and female twins may need closer scrutiny and perhaps sex-specific reanalysis.

Muscle Tissue and Training

Muscle tissue is the substrate of strength and movement and can be influenced by training. Hence, it is logical to inquire into the genotypic component of variance in muscle tissue characteristics. Komi, Viitasalo, Havu, Thorstensson, Sjödin, and Karlsson (1977) reported that MZ twins (13 pairs) were virtually identical in the percentage of Type I (slow twitch) fibers, while DZ twins (16 pairs) were quite variable. The estimated heritability approached unity (0.96).

Table 31. Intrapair correlations for learning curve parameters in Polish twins

| | Level of learning | | | | Rate of learning | | | |
| | Males | | Females | | Males | | Females | |
Motor task	MZ	DZ	MZ	DZ	MZ	DZ	MZ	DZ
Plate tapping, R	.92	.70	.57	.14	.93	.29	.49	.09
L	.93	.60	.60	.05	.65	.37	.64	.13
One foot tapping, R	.93	.53	.54	−.22	.53	.74	.54	.07
L	.95	.39	.65	−.14	.90	.36	.53	.11
Mirror tracing, errors	.70	.42	.87	.31	.44	−.13	.88	.31
time	.76	.75	.80	.41	.70	.11	.20	.41
Ball toss, R	.67	.71	.33	.20	.61	.48	.91	.26
L	.72	.26	.54	−.24	.43	.60	.68	−.10

Note. Data compiled from Skład (1975); n male MZ = 14, DZ = 11; n female MZ = 10, DZ = 11.

More recently, however, Lortie, Simoneau, Boulay, and Bouchard (1986) noted only moderate genotype dependency in the percentage of Type I fibers in MZ twins (35 pairs, intraclass correlation, 0.33). The percentage of Type IIa fibers showed little similarity, while the percentage of Type IIb fibers showed moderate similarity in genetically related individuals.

Muscle size is also influenced by genetic factors. Radiographic analysis of calf musculature indicated correlations of 0.56 for male and 0.63 for female sibling pairs of preschool age (Hewitt, 1957), and high intraclass correlations of 0.83 for male and 0.85 for female MZ twins 12 to 13 years of age (Hoshi, Ashizawa, Kouchi, & Koyama, 1982).

In contrast to fiber types and muscle size, the observations of Komi et al. (1977) showed no significant genetic variation in activities of muscle ATPases, CPK, myokinase, phosphorylase, and lactate dehydrogenase as indicated by small differences between MZ and DZ twin pairs. In an analysis of mitochondria in 11 MZ and 6 DZ twin pairs, Howald (1976) reported no gene-associated variation in mitochondrial density, the ratio of mitochondrial volume to myofibril volume, and the internal and external surface densities of mitochondria. Furthermore, activities of hexokinase and SDH were identical in both sets of twins, while muscle glyceraldehyde-3-phosphate dehydrogenase and 3-hydroxyacyll-CoA-dehydrogenase were significantly more variable among DZ twins than among MZ twins.

These observations may in part reflect training differences. With one member of an MZ twin pair (n = 7 pairs) participating in a 23-week endurance training program while the other maintained his usual activity, Howald (1976) noted significant increases in selected extra- and intramitochondrial enzymes. SDH (intra) and hexokinase (extra) activities, for example, increased by 28% and 17%, respectively, with training in one member of the MZ pair compared to the other. Also working with MZ twins, Komi, Viitasalo, Rauramaa, and Vihko (1978) noted training associated variation during a 12-week isometric strength training program. With training, the enzyme activities of malic dehydrogenase, succinate dehydrogenase, and creatine kinase were lowered in the trained leg compared to the untrained leg of the experimental twin subject, and compared to the values for the enzyme activities in the control member of the MZ twin pair. The trained leg of the experimental twin gained 20% in strength and the untrained leg gained 11%. In contrast, the control twin did not show any change in strength.

Regular training apparently does not influence the distribution of muscle fiber types. However, strength, fiber size and area, and the relative area of a muscle composed of Type I and Type II fibers may change with training. Some evidence, though limited at present, also suggests a possible transformation of fiber types with training (see Malina, 1983, 1985). Thus, although genetic contributions to muscle tissue fiber composition and size are significant, the evidence emphasizes a significant role for physical training in modifying size and metabolic capacity.

There is a need for comprehensive assessment of the relationship between muscle fibers and function. The percentage of Type I fibers had a moderate relationship with time required to reach a specific force level in an isometric bilateral leg extension movement (r = + 0.48, Viitasalo & Komi, 1978), and with the isometric force of the quadriceps muscle group in males (r = -0.55, Komi & Karlsson, 1978). However, in samples of MZ and DZ twins, the force-time measurement and quadriceps force had a relatively small genetic component as indicated by nonsignificant intrapair variance ratios (Komi et al., 1973; Komi & Karlsson, 1979). On the other hand, maximal muscular power, as derived from a running velocity test, had no relationship to the percentage of Type I fibers; yet this test of muscular power had a very high estimated heritability (Komi et al., 1973; Komi & Karlsson, 1978, 1979).

In a developmental perspective, children with idiopathic late walking (mean age, 19 months) show reduced muscle fiber size, especially of Type II fibers (Lundberg, Eriksson, & Jansson, 1979), while those with celiac disease (mean age, 14 months) have a reduced percentage of Type I fibers and poor gross motor development (Lundberg, Erikkson, & Mellgren, 1979). After dietary treatment in the latter, the percentage of Type I fibers increased, as did fiber diameter and gross motor development. Physical inactivity may play a role in such conditions, as similar changes occur with atrophy due to inactivity. On the other hand, the influence of regular physical activity on muscle tissue morphometry and function is significant.

Perceptual and Perceptual-Motor Characteristics

Individual differences in perceptual characteristics such as spatial abilities, perceptual speed, perception of direction, and perceptual integration have a significant genetic component (McGee, 1979; Rose, Miller, Dumon-Driscoll, & Evans, 1979; Vandenberg, 1962, 1966). The role of such perceptual characteristics in skillful motor performance, though often suggested, has not been fully investigated. Kolakowski and Malina (1974), for example, reported a significant relationship between spatial ability (Primary Mental Abilities Test) and baseball throwing accuracy in boys 14 to 16 years of age. Several accuracy scoring systems were used, and only accuracy scored in the vertical plane (in contrast to the horizontal plane and concentric circles) had a significant association with spatial ability. This relationship between spatial ability and vertical accuracy may be significant in performance since vertical deviation on an upright target would translate into skill and judgment with respect to distance if the target were laid on the ground.

Reaction time is another characteristic often deemed important in motor performance. However, results of twin studies are variable for the genetic component of reaction time. Vandenberg (1962) noted a low heritability (0.22) for reaction time to a light stimulus in 43 MZ and 29 same-sex DZ pairs of high school twins, while Skład (1973) reported a slightly higher heritability (0.55) for a similar task in 14 MZ and 14 same-sex DZ pairs 8 to 15 years of age. On the other hand, Komi et al. (1973) reported high heritabilities for visual reaction time (0.86) and patellar reflex time (0.97) in eight pairs of MZ and seven pairs of DZ male twins. Among female twins, however, seven pairs of DZ and seven pairs of MZ twins were equally variable in these two measurements. Intrapair variances between male MZ and DZ twins and female MZ and DZ twins for ulnar nerve conduction velocity did not differ significantly.

There are also parent-offspring similarities in reaction time. Results for a rural Polish sample are shown in Table 32. All correlations were moderate to moderately high, and parent-daughter correlations tended to be higher than parent-son values.

Concluding Comments

Estimates of the genotypic component of variance in motor development and performance vary among studies, tasks, and types of genetically related individuals considered. Most of the data are derived from twins, but the results are unequal in quantitative value. Many studies are limited to males or females, while others combine the sexes. Some data from twins also suggest higher heritabilities for males than females. The more limited sibling data suggest a similar trend: Brothers tend to resemble each other more than sisters in strength and motor tasks. This would in turn suggest sex-influences in estimated heritabilities of performance items. Perhaps the simplest explanation would be an environmental covariation that differs with respect to sex. If one brother is proficient in motor skills, the other brother is more likely to be proficient in similar tasks, perhaps due to social or familial pressures or the more ready acceptance of males in sports. On the other hand, the physical activity pursuits of one sister are less likely to influence the pursuits of the other.

Quantitative genetic studies have potential limitations. As is the case for most biological traits, the evidence indicates a moderate heritability for many motor tasks with an unknown environmental effect. And the general model used to estimate the contribution of genetic variation to multifactorial traits assumes that the environmental interaction is not significant (Bouchard & Malina, 1983a). Little effort has been made to specify the environmental sources of variation in motor performance as well as the genotype-environment interaction. There is thus a need to quantify specific environmental factors that influence motor performance, for example, training, level of habitual physical activity, and the shared sibling and familial environment. These are nonrandom environmental sources of variance and may lead to inflated heritability estimates, as was recently demonstrated with the use of path analysis in the evaluation of familial resemblance in six neuromuscular traits (Devor &

Table 32. Parent-offspring correlations for reaction time in a rural Polish sample

Stimulus	F-S (n = 69-79)	M-S (n = 89-107)	MP-S (n = 51-61)	F-D (n = 56-64)	M-D (n = 77-89)	MP-D (n = 41-48)
Optical	.59	.51	.63	.70	.57	.73
Acoustic	.54	.44	.61	.64	.66	.75
Tactile	.55	.54	.65	.64	.57	.72

Note. Data compiled from Wolański and Kasprzak (1979); age of siblings 6 to 39 years. All correlations $p < .01$.

Crawford, 1984). The traits varied in the degree of parent-offspring transmission but all six showed substantial residual sibling resemblance due to common environmental effects.

Finally, in virtually all studies the outcome of performance is measured and subsequently analyzed in terms of genotypic and environmental sources of variance. The underlying processes of performance are not considered. There is thus a need to systematically investigate the biochemical, physiological, and biomechanical correlates of performance in a genetic context.

References

Asmussen, E. (1968). The neuromuscular system and exercise. In H.B. Falls (Ed.), *Exercise physiology* (pp. 3-42). New York: Academic Press.

Baldwin, J.C., & Damon, A. (1973). Some genetic traits in Solomon Island populations. V. Assortative mating, with special reference to skin colour. *American Journal of Physical Anthropology*, **39**, 195-201.

Bayley, N. (1965). Comparisons of mental and motor test scores for ages 1-15 months by sex, birth order, race, geographical location, and education of parents. *Child Development*, **36**, 379-411.

Bayley, N. (1969). *Manual for the Bayley scales of infant development*. New York: Psychological Corporation.

Boileau, R.A., & Lohman, T.G. (1977). The measurement of human physique and its effect on physical performance. *Orthopedic Clinics of North America*, **8**, 563-581.

Bossik, L.J. (1934). K voprosu o roli nasledsvennosti i sredi v fiziologii i patologii detskova vovrasta. *Trudy Medytsinsko-Biologicheskovo Instituta*, **3**, 33-56 (as cited by Skład, 1972 and 1973).

Bouchard, C., & Lortie, G. (1984). Heredity and endurance performance. *Sports Medicine*, **1**, 38-64.

Bouchard, C., & Malina, R.M. (1983a). Genetics for the sport scientist: Selected methodological considerations. *Exercise and Sport Sciences Reviews*, **11**, 275-305.

Bouchard, C., & Malina, R.M. (1983b). Genetics of physiological fitness and motor performance. *Exercise and Sport Sciences Reviews*, **11**, 306-339.

Brody, D. (1937). Twin resemblances in mechanical ability, with reference to the effects of practice on performance. *Child Development*, **8**, 207-216.

Clarke, H.H. (1967). *Application of measurement to health and physical education*. Englewood Cliffs, NJ: Prentice-Hall.

Cook, C.F., & Broadhead, G.D. (1984). Motor performance of pre-school twins and singletons. *The Physical Educator*, **41**(1), 16-20.

Cratty, B.J. (1960). A comparison of fathers and sons in physical ability. *Research Quarterly*, **31**, 12-15.

Dales, R.J. (1969). Motor and language development of twins during the first three years. *Journal of Genetic Psychology*, **114**, 263-271.

Devor, E.J., & Crawford, M.H. (1984). Family resemblance for neuromuscular performance in a Kansas Mennonite community. *American Journal of Physical Anthropology*, **64**, 289-296.

Engström, L.M., & Fischbein, S. (1977). Physical capacity in twins. *Acta Geneticae Medicae et Gemellologiae*, **26**, 159-165.

Freedman, D.G. (1974). *Human infancy: An evolutionary perspective*. New York: Wiley.

Freedman, D.G., & Keller, B. (1963). Inheritance of behavior in infants. *Science*, **140**, 196-198.

Fujikura, T., & Froehlich, L.A. (1974). Mental and motor development in monozygotic co-twins with dissimilar birth weights. *Pediatrics*, **53**, 884-889.

Garrison, R.J., Anderson, V.E., & Reed, S.C. (1968). Assortative marriage. *Eugenics Quarterly*, **15**, 113-127.

Gesell, A. (1954). The ontogenesis of infant behavior. In L. Carmichael (Ed.), *Manual of child psychology* (pp. 335-373). New York: Wiley.

Gesell, A., & Thompson, H. (1929). Learning and growth in identical infant twins: An experimental study by the method of co-twin control. *Genetic Psychology Monographs*, **6**, 1-124.

Gifford, S., Murawski, B.J., Brazelton, T.B., & Young, G.C. (1966). Differences in individual development within a pair of identical twins. *International Journal of Psycho-analysis*, **47**, 261-268.

Harrison, G.A., Gibson, J.B., & Hiorns, R.W. (1976). Assortative marriage for psychometric, personality, and anthropometric variation in a group of Oxfordshire villages. *Journal of Biosocial Science*, **8**, 145-153.

Hewitt, D. (1957). Sib resemblance in bone, muscle and fat measurements of the human calf. *Annals of Human Genetics*, **22**, 213-221.

Hilgard, J.R. (1933). The effect of early and delayed practice on memory and motor performance studied by the method of co-twin control. *Genetic Psychology Monographs*, **14**, 493-567.

Hoshi, H., Ashizawa, K., Kouchi, M., & Koyama, C. (1982). On the intrapair similarity of Japanese monozygotic twins in some somatological traits. *Okajimas Folia Anatomica Japonica,* **58**, 675-686.

Hoshi, H., Takahashi, Ch., Ashizawa, K., & Kouchi, M. (1980). Revised percent deviation and coefficient of similarity: Study for classification of physique based on twin method. Report 1. *Journal of the Anthropological Society of Nippon*, **88**, 9-24.

Howald, H. (1976). Ultrastructure and biochemical function of skeletal muscle in twins. *Annals of Human Biology*, **3**, 455-462.

Ishidoya, Y. (1957). Sportfähigkeit der Zwillinge. *Acta Geneticae Medicae et Gemellologiae*, **6**, 321-326.

Johnson, R.C., DeFries, J.C., Wilson, L.R., McClearn, G.E., Vandenberg, S.G., Ashton, G.C., Mi, M.P., & Rashad, M.N. (1976). Assortative marriage for specific cognitive abilities in two ethnic groups. *Human Biology*, **48**, 343-352.

Kimura, K. (1956). The study on physical ability of children and youths: On twins in Osaka City. *Jinrui-gaku Zasshi* (Anthropological Society of Nippon), **64**, 172-196.

Kolakowski, D., & Malina, R.M. (1974). Spatial ability, throwing accuracy, and man's hunting heritage. *Nature*, **251**, 410-412.

Komi, P.V., & Karlsson, J. (1978). Skeletal muscle fibre types, enzyme activities and physical performance in young males and females. *Acta Physiologica Scandinavica*, **103**, 210-218.

Komi, P.V., & Karlsson, J. (1979). Physical performance, skeletal muscle enzyme activities, and fibre types in monozygous and dizygous twins of both sexes. *Acta Physiologica Scandinavica* (Supplement 462).

Komi, P.V., Klissouras V., & Karvinen, E. (1973). Genetic variation in neuro-muscular performance. *Internationale Zeitschrift für angewandte Physiologie*, **31**, 289-304.

Komi, P.V., Viitasalo, J.H.T., Havu, M., Thorstensson, A., Sjödin, B., & Karlsson, J. (1977). Skeletal muscle fibres and muscle enzyme activities in monozygous and dizygous twins of both sexes. *Acta Physiologica Scandinavica*, **100**, 385-392.

Komi, P.V., Viitasalo, J.T., Rauramaa, R., & Vihko, V. (1978). Effect of isometric strength training on mechanical, electrical, and metabolic aspects of muscle function. *European Journal of Applied Physiology*, **40**, 45-55.

Kovař, R. (1974). *Prispevek ke studiu geneticke podminenosti lidske motoriky.* Doctoral dissertation, Charles University, Prague.

Kovař, R. (1981a). *Human variation in motor abilities and its genetic analysis*. Prague: Charles University.

Kovař, R. (1981b). Sledovani podobnosti mezi rodici a jejich potomky v nekterych motorickych projevech. *Teorie a Praxe Telesne Vychovy*, **29**, 93-98.

Larson, L.A., & Yocom, R.D. (1951). *Measurement and evaluation in physical, health and recreation education*. St. Louis: Mosby.

Lortie, G., Simoneau, J.-A., Boulay, M., & Bouchard, C. (1986). Muscle fiber type composition and enzyme activities in brothers and monozygotic twins. In R.M. Malina & C. Bouchard (Eds.), *Sport and human genetics* (pp. 147-153). Champaign, IL: Human Kinetics.

Lundberg, A., Eriksson, B.O., & Jansson, G. (1979). Muscle abnormalities in coeliac disease: Studies on gross motor development and muscle fibre composition, size and metabolic substrates. *European Journal of Pediatrics*, **130**, 93-103.

Lundberg, A., Eriksson, B.O., & Mellgren, G. (1979). Metabolic substrates, muscle fibre composition and fibre size in late walking and normal children. *European Journal of Pediatrics*, **130**, 79-92.

McGee, M.G. (1979). Human spatial abilities: Psychometric studies and environmental, genetic, hormonal, and neurological influences. *Psychological Bulletin*, **86**, 889-918.

McGraw, M.B. (1935). *Growth: A study of Johnny and Jimmy*. New York: Appleton-Century.

McNemar, Q. (1933). Twin resemblances in motor skills, and the effect of practice thereon. *Pedagogical Seminary and Journal of Genetic Psychology*, **42**, 70-99.

Malina, R.M. (1975). Anthropometric correlates of strength and motor performance. *Exercise and Sport Sciences Reviews*, **3**, 249-274.

Malina, R.M. (1982). Motor development in the early years. In S.G. Moore & C.R. Cooper (Eds.), *The young child: Reviews of research* (Vol. 3, pp. 211-229). Washington, DC: National Association for the Education of Young Children.

Malina, R.M. (1985). Human growth, maturation, and regular physical activity. *Acta Medica Auxologica*, **15**, 5-23.

Malina, R.M. (In press). Growth of muscle tissue and muscle mass. In F. Falkner & J.M. Tanner (Eds.), *Human growth. Volume 2. Postnatal growth*. New York: Plenum.

Malina, R.M., & Buschang, P.H. (1985). Growth, strength and motor performance of Zapotec children, Oaxaca, Mexico. *Human Biology*, **57**, 163-181.

Malina, R.M., & Little, B.B. (1985). Sibling similarities in the strength and motor performance of undernourished children of school age. (Submitted for publication).

Malina, R.M., & Mueller, W.H. (1981). Genetic and environmental influences on the strength and motor performance of Philadelphia school children. *Human Biology*, **53**, 163-179.

Malina, R.M., Selby, H.A., Buschang, P.H., Aronson, W.L., & Little, B.B. (1983). Assortative mating for phenotypic characteristics in a Zapotec community in Oaxaca, Mexico. *Journal of Biosocial Science*, **15**, 273-280.

Marisi, D.Q. (1977). Genetic and extragenetic variance in motor performance. *Acta Geneticae Medicae et Gemellologiae*, **23**, 197-204.

Mattson, M.L. (1933). The relation between complexity of the habit to be acquired and the form of the learning curve in young children. *Genetic Psychology Monographs*, **13**, 299-398.

Mirenva, A.N. (1935). Psychomotor education and the general development of preschool children: Experiments with twin controls. *Pedagogical Seminary and Journal of Genetic Psychology*, **46**, 433-454.

Mizuno, T. (1956). Similarity of physique, muscular strength and motor ability in identical twins. *Bulletin of the Faculty of Education, Tokyo University*, **1**, 136-157.

Montoye, H.J., Metzner, H.L., & Keller, J.K. (1975). Familial aggregation of strength and heart rate response to exercise. *Human Biology*, **47**, 17-36.

Roberts, D.F. (1977). Assortative mating in man: Husband/wife correlations in physical characteristics. *Bulletin of the Eugenics Society* (Supplement 2).

Rose, R.J., Miller, J.Z., Dumon-Driscoll, M., & Evans, M.M. (1979). Twin-family studies of perceptual speed ability. *Behavioral Genetics*, **9**, 71-86.

Skład, M. (1972). Similarity of movements in twins. *Wychowanie Fizycznie i Sport*, **16**, 119-141.

Skład, M. (1973). Rozwój fizyczny i motoryczność blizniat. *Materialy i Prace Antropologiczne*, **85**, 3-102.

Skład, M. (1975). The genetic determination of the rate of learning motor skills. *Studies in Physical Anthropology*, **1**, 3-19.

Smith, N.W. (1976). Twin studies and heritability. *Human Development*, **19**, 65-68.

Spuhler, J.N. (1968). Assortative mating with respect to physical characteristics. *Eugenics Quarterly*, **15**, 128-140.

Susanne, C. (1967). Contribution a l'etude de l'assortiment matrimonial dans un echantillon de la population Belge. *Bulletin de la Societe Royale Belge d'Anthropologie et de Prehistoire*, **78**, 147-196.

Szopa, J. (1982). Familial studies on genetic determination of some manifestations of muscular strength in man. *Genetica Polonica*, **23**, 65-79.

Szopa, J. (1983). Zmienność oraz genetyczne uwarunkowania niektórych przejawów sily miesni u człowieka wyniki badań rodzinnych. *Materialy i Prace Antropologiczne*, **103**,131-154.

Vandenberg, S.G. (1962). The hereditary abilities study: Hereditary components in a psychological test battery. *American Journal of Human Genetics*, **14**, 220-237.

Vandenberg, S.G. (1966). Contributions of twin research to psychology. *Psychological Bulletin*, **66**, 327-352.

Venerando, A., & Milani-Comparetti, M. (1970). Twin studies in sport and physical performance. *Acta Geneticae Medicae et Gemellologiae*, **19**, 80-82.

Verschuef, O. von (1925). Die Wirkung der Umwelt auf die anthropologischen Merkmale nach Untersuchungen an eineiigen Zwillingen. *Archiv für Rassen-und Gessellschaft-Biologie*, **17**, 149-164.

Verschuer, O. von (1927). Studien an 102 eineiigen und 45 gleichgeschlechtlichen zweieiigen Zwillings und an 2 Drillingspaaren. *Ergebnisse der Inneren Medizin und Kinderheilkunde*, **31**, 35-120 (as cited by Skład, 1972 and 1973).

Vickers, V.S., Poyntz, L., & Baum, M.P. (1942). The Brace scale used with young children. *Research Quarterly*, **13**, 299-308.

Viitasalo, J.T., & Komi, P.V. (1978). Force-time characteristics and fiber composition in human leg extensor muscles. *European Journal of Applied Physiology*, **40**, 7-15.

Weiss, V. (1977). Die Heritabilitäten sportlicher Tests, berechnet aus den Leistungen zehnjähriger Zwillingspaare. *Arztliche Jugendkunde*, **68**, 167-172.

Wilde, G.J.S. (1970). An experimental study of mutual behaviour imitation and person perception in MZ and DZ twins. *Acta Geneticae Medicae et Gemellologiae*, **19**, 273-279.

Williams, H.G. (1983). *Perceptual and motor development*. Englewood Cliffs, NJ: Prentice-Hall.

Williams, L.R.T., & Gross, J.B. (1980). Heritability of motor skill. *Acta Geneticae Medicae et Gemellologiae*, **29**, 127-136.

Williams, L.R.T., & Heartfield, V. (1973). Heritability of a gross motor balance task. *Research Quarterly*, **44**, 109-112.

Wilson, R.S., & Harpring, E.B. (1972). Mental and motor development in infant twins. *Developmental Psychology*, **7**, 277-287.

Wolański, N. (1973). Assortative mating in the Polish rural populations. *Studies in Human Ecology*, **1**, 182-188.

Wolański, N., & Kasprzak, E. (1979). Similarity in some physiological, biochemical and psychomotor traits between parents and 2-45 years old offspring. *Studies in Human Ecology*, **3**, 85-131.

Wolański, N., Tomonari, K., & Siniarska, A. (1980). Genetics and the motor development of man. *Human Ecology and Race Hygiene*, **46**, 169-191.

3

Genetics of Aerobic Power and Capacity

Claude Bouchard
LAVAL UNIVERSITY
QUEBEC, CANADA

In recent reviews we have examined the question of genetic methods in sport studies (Bouchard & Malina, 1983), the genetics of physiological fitness and motor performance (Bouchard & Malina, 1983a), the importance of heredity in endurance performance (Bouchard & Lortie, 1984), the general topic of inheritance and the Olympic athletes (Bouchard & Malina, 1984), and the role of the genotype in body fat and fat cell metabolism (Després & Bouchard, 1984). In this paper the influences of biological inheritance on maximal aerobic power and capacity are summarized and discussed. The phenotypes of maximal aerobic power (i.e., the peak aerobic metabolic rate of the working human organism), and of maximal aerobic capacity or of its estimators (i.e., the capacity for prolonged aerobic work), are considered. Moreover, selected cardiac, skeletal muscle, and substrate determinants of aerobic exercise are summarized.

The effects of genes on a given phenotype can occur in at least three different ways: (a) by their contribution to trait(s) correlated with the phenotype, (b) by the heritability, that is, the mean genetic effect, of the phenotype for the given population, and (c) by the importance of the genotype dependence of the adaptive response to training or other lifestyle components (Bouchard & Lortie, 1984; Bouchard & Malina, 1984). In this paper only the last two of these effects are considered. However, there is ample evidence that genes

Thanks are expressed to Drs. Claude Allard, Marcel R. Boulay, Fernand Landry, Germain Thériault, and Angelo Tremblay, and to Claude Leblanc, who have been closely associated with these studies over the years. Thanks are also expressed to the graduate students who contributed much to the research program and whose works have been quoted here, and to research associates and technicians who participated to this research program over the years. This research is supported by grants from FCAC-Quebec (EQ-1330) and NSERC of Canada (G-0850 and A-8150).

also affect performances in aerobic exercise via their action on traits which covary with these phenotypes. For instance, long-distance runners are generally lean individuals (Wilmore, Brown, & Davis, 1977) with thin subcutaneous fat deposits (Carter, 1982) and small mean fat cell diameter (Després, Savard, Tremblay, & Bouchard, 1983). They rate high in ectomorphy, low in mesomorphy, and very low in endomorphy (Carter, 1978) and generally have small size, short legs, and narrow shoulders (Tanner, 1964). These covariates of aerobic performances should therefore be given proper consideration when discussing human variation in aerobic power and capacity even though they will not be considered in this review.

The Multifactorial Model

We have reviewed elsewhere the basic model of quantitative genetics applicable to the study of phenotypes influenced by inherited factors, lifestyle components, and their interaction (Bouchard & Malina, 1983). Such phenotypes, including maximal aerobic power and capacity and their respective determinants, are appropriately referred to as multifactorial phenotypes. Briefly, the total phenotypic variance for such traits can be partitioned as follows:

$$V_P = V_G + V_E + V_{GxE} + e$$

In this equation, V represents the variance, P is the phenotype, G is the genetic component of the variance, E is the environmental effect (i.e., the nongenetic), GxE represents the interaction effects between the genotype and environmental factors, and e is the random error component.

The Phenotypes

Brief mention is warranted about the measurement of maximal aerobic power, maximal aerobic capacity, and some of its estimators and their determinants. Maximal aerobic power (MAP), or maximal oxygen uptake, is measured under laboratory conditions. In the experiments from our laboratory, MAP was obtained on the cycle ergometer unless otherwise specified. MAP/kg is a highly reproducible measurement with an intraclass reliability coefficient of about 0.96 (Prud'homme et al., 1984a). In the case of MAP/kg therefore, the e component is probably quite small and is approximately 5% in our laboratory.

There is no commonly accepted procedure to measure maximal aerobic capacity (MAC), that is, the total capacity for prolonged aerobic work, and little has been reported on the subject. However, there is need for a test to estimate reliably the organism's capacity to perform for a reasonably long period of time under aerobic conditions. Such a test was recently developed in our laboratory. Briefly, the test requires that the subject work for 90 minutes on a cycle ergometer at the highest sustainable intensity. MAC is expressed as the total work output (kJ) in 90 minutes. MAC/kg is highly reproducible in both sexes (Boulay et al., 1984), with an intraclass reliability coefficient of 0.98.

Ventilatory thresholds are believed to be well correlated with performance in prolonged aerobic exercise. Thus, data from our laboratory indicate that the first (VT-1) and the second (VT-2) nonlinear increases in ventilation are correlated with MAC/kg with a coefficient $\geqslant 0.82$ (Boulay, Lortie, Simoneau, Leblanc, & Bouchard, unpublished data). Moreover, VT-1 and VT-2 expressed in ml O_2/kg•min^{-1} (i.e., VT-1/kg and VT-2/kg) are reproducible measurements with intraclass coefficients of about 0.9 and above (Prud'homme et al., 1984a).

Submaximal power output in relative steady state at a heart rate of 150 beats per minute (PWC$_{150}$) was also considered in some of our experiments on heredity and aerobic metabolism. PWC$_{150}$/kg body weight measured on the cycle ergometer is a highly reproducible variable with a reliability coefficient $\geqslant 0.96$ (Bouchard et al., 1984). PWC$_{170}$/kg also correlates well with MAP/kg and MAC/kg (Boulay et al., unpublished data).

Other phenotypes in this review include selected determinants of aerobic power and capacity. Thus, heart dimensions from echographic measurements, skeletal muscle characteristics (fiber type, oxidative enzyme activities), and substrate availability (e.g., basal and catecholamine stimulated fat cell lipolysis) are also considered. These measurements are routinely obtained with satisfactory reliability in our laboratory (Després, Bouchard, Bukowiecki, Savard, & Lupien, 1983a; Simoneau, Lortie, Boulay, Thibault, & Bouchard, unpublished data; Landry, Bouchard, & Dumesnil, unpublished data).

It thus seems reasonable to conclude that the phenotypes discussed in this review are generally measured in the laboratory with a high degree of repeatability. One could therefore feel reasonably safe in assuming that e (the error component) is equal to zero in the model and consider only the E, G, and GxE effects on the phenotypes.

Besides the random error component (e), one must deal with other unwarranted effects that are assumed not to be present in the model. For instance, it is commonly assumed that there are no age and gender effects on the phenotype under consideration. But if there are, the attempt to estimate the E, G, and GxE effects is then destined to failure. In other words, precaution must be taken to control experimentally or statistically for the concomitant variables associated with the P under consideration in order to generate appropriate conditions for a valid assessment of the components of the model. In addition, whenever appropriate, one should also take note of strong indications of skewness and kurtosis of distribution which reveal serious deviation from normality, as it may affect the population estimates.

The E Effects

Within the model described in this paper, the E effect is approximately equivalent to the mean training effect in the population for the phenotype. The reduction of the E component to training is justified in this case for two major reasons. First, MAP and MAC are known to be influenced by training, and no other lifestyle factor can reasonably be claimed to cause major variation in a normal free-living population of well nourished individuals. Second, it has already been shown that correlations between estimated MAP and socioeconomic status, smoking, and other nongenetic and nontraining factors are quite low (Lortie, Bouchard, Simoneau, & Leblanc, 1983).

To obtain acceptable estimates of the E effects in comparison with the G and GxE components, it is appropriate to rely primarily on experiments specially designed to quantify all three components of the system in an effort to establish their relative contribution. Such experiments have been undertaken in our laboratory during the past few years. These experiments were carried out on nuclear families with biological children or adopted children, dizygotic (DZ) and monozygotic (MZ) twins, first-degree cousins, uncles (aunts), and nephews (nieces), and unrelated individuals living together or apart. The experiments required to estimate the effect of E were done with subjects ascertained as sedentary in the months prior to the study and with no history of regular participation in physical activities or training. These subjects were selected in the age range of approximately 18 to 30 years, to reduce as much as possible the influence of age. There were both males and females, however, as we were interested in a potential gender effect in the causal sources of variation of MAP or MAC.

Maximal Aerobic Power and Capacity

Under the influence of an endurance training program, the organism adapts progressively over a period of several months. Both the oxygen delivery system (heart, respiration, circulation) and the oxygen utilization system (muscle blood circulation, muscle substrate availability, muscle ATP replenishment machinery, etc.) increase their capacity and power at various rates with training. Some of these adaptive changes have been summarized by Saltin, Henriksson, Nygaard, Andersen, and Jansson (1977).

Although much of the research indicates that maximal aerobic power can be improved with training, there are also clear indications that trainability is limited. The plasticity of maximal aerobic power remains quite high, however. This was clearly demonstrated years ago in the classic bedrest training study of Saltin et al. (1968) (see also Saltin, 1972). Five subjects (3 untrained) were tested for maximal aerobic power and related circulatory variables and then put to bed for 20 days. After the bedrest, the subjects were retested and they then began an intensive training period of 55 days. In the untrained subjects the average reduction in maximal aerobic power was 30%. After 55 days of training these subjects showed an increase of 33% over the pre-bedrest level. The same training program caused no change in maximal heart rate but a 16% increase in maximal cardiac output, maximal stroke volume, and maximal O_2 arteriovenous difference.

It is commonly suggested that maximal aerobic power will easily increase by 10 to 20% under the influence of a training program stressing primarily the aerobic processes. Klissouras (1972) has suggested from observation on pairs of MZ twins that maximal trainability is probably around 40% of the initial value. More recent data submitted by Hickson, Bomze, and Holloszy (1977) suggest that maximal aerobic power in sedentary subjects increases linearly during a 10-week intensive training program. These authors report an average weekly increase of about 0.12 1 of $O_2 \cdot min^{-1}$ in maximal aerobic power. The total improvement for 8 subjects over 10 weeks was 39%. On the other hand, some data suggest that the capacity of aerobic work metabolism generally is more trainable than the power of the system.

In one experiment designed to estimate E for the aerobic power and capacity phenotypes, 24 individuals (13 women and 11 men) were subjected to a 20-week cycle ergometer endurance training program (Lortie et al., 1984). Subjects trained initially 4 times per week and increased this to 5 times a week. Each session lasted 40 to 45 minutes, starting at 60% and increasing to 85% of the heart rate reserve. Table 1 illustrates the training effects of this aerobic program on MAP/kg, MAC/kg, VT-1/kg, and VT-2/kg. For these sedentary subjects, the standardized 20-week aerobic training program results in a 33% increase in MAP/kg but a 51% increase in MAC/kg. Although the response of maximal aerobic power to endurance training was similar in both sexes, men improved 50% more than women in the maximal aerobic capacity test. In this experiment, VT-1/kg and VT-2/kg improved by more than 30%, that is, by as much as maximal aerobic power but less than aerobic capacity. The relationship between training gains and the pretraining scores reached about −0.5—a common variance of 25%.

Determinants of MAP and MAC

Recent reviews (Keul, Dickhuth, Lehmann, & Staiger, 1982; Rost & Hollmann, 1983) have indicated that the endurance athlete has a large heart size, bradycardia at rest, a larger stroke volume and cardiac output during maximal exercise than other individuals, a reduced heart rate for the same submaximal work load with little difference in maximal heart rate, a decreased arterial pressure at rest and for a given work load, and an increased arteriovenous oxygen difference at exhaustion. It is fairly well established that some of these characteristics can be partly developed as a result of endurance training (Blomqvist & Saltin, 1983; Keul et al., 1982), but doubts have been expressed about the effects of training on the enlargement of the heart and hypertrophy of the myocardium.

Table 2 summarizes the findings of one of our recent studies concerning the effects of aerobic training on the echographic heart dimensions (Landry et al., unpublished data). Sedentary subjects took part in the same 20-week training program as described above (see Table 1). Although mean changes with training were significant and in the predicted direction, actual changes in heart dimensions were generally small. In other words, these sedentary subjects experienced a 33% mean increase in MAP/kg with modest changes in heart dimensions ranging from a mean of 2% to 19%.

The fiber composition and metabolic characteristics of the skeletal muscle are thought to have a significant influence on MAP and MAC. Thus, athletes in endurance sports have a higher proportion of type I fibers than nonathletes in the muscles used during training (Costill, Fink, & Pollock, 1976; Gollnick, Armstrong, Saubert, Piehl, & Saltin, 1972). But results are not always as clear when one looks into the relationship between the percentage of type I muscle fibers and various criteria of aerobic performance. For instance, the correlation between the percentage of type I fibers and MAP/kg ranges from about zero (Rusko, Rahkila, & Karvinen, 1980) to more than 0.6 (Foster, Costill,

Table 1. The influence of a 20-week aerobic training program on measurements of aerobic performance in 24 subjects ascertained as sedentary[a]

Variable	Pretraining M (SD)	Posttraining M (SD)	Individual changes in % M (SD)	Range in %	Correlation of % gain with pretraining value (r)
MAP/kg (ml O_2/kg·min^{-1})	37 (7)	48 (7)**	33 (15)	5 to 88	−0.6**
MAC/kg (J/kg in 90 min)	9.1 (1.9)	13.6 (2.7)**	51 (22)	16 to 97	−0.5*
VT-1/kg (ml O_2/kg·min^{-1})	22 (5)	28 (6)**	31 (28)	−12 to 107	−0.5*
VT-2/kg (ml O_2/kg·min^{-1})	30 (6)	40 (7)**	36 (16)	3 to 79	−0.5*

[a]From Lortie et al. (1984) and unpublished data.
*$p \leqslant 0.05$; **$p \leqslant 0.01$.

Table 2. The influence of a 20-week aerobic training program on echographic heart dimensions in 20 subjects ascertained as sedentary[a]

Variable	Pretraining M (SD)	Posttraining M (SD)	Individual changes in % M (SD)	Individual changes in % Range in %	Correlation of % gain with pretraining value (r)
Left ventricle internal diameter (mm)	47.1 (3.8)	47.9 (3.3)*	2 (3)	−5 to 10	−0.5*
Left ventricle posterior wall thickness (mm)	7.5 (1.2)	8.3 (1.3)**	10 (13)	−20 to 39	−0.4
Interventricular septum thickness (mm)	8.1 (1.2)	8.9 (1.1)**	11 (13)	−11 to 43	−0.5*
Left ventricle volume/BSA[b] (ml/m²)	64 (12)	68 (11)*	6 (10)	−14 to 31	−0.5*
Left ventricle mass/BSA (g/m²)	91 (19)	106 (22)**	19 (19)	−17 to 56	−0.3

[a]Adapted from Landry et al. (unpublished data). All measurements in diastole.
[b]BSA = body surface area.
*p ⩽ 0.05; **p ⩽ 0.01.

Daniels, & Fink, 1978; Ivy, Withers, Van Handel, Elger, & Costill, 1980). On the other hand, correlations between the percentage of type I fibers and the onset of blood lactate accumulation vary from 0.52 to 0.75 (Komi, Ito, Sjödin, Wallenstein, & Karlsson, 1981; Tesch, Daniels, & Sharp, 1982). The correlation between performance times for running distances from 1 to 6 miles and the percentage of type I fibers is 0.54 (Foster et al., 1978), while that for the average speed during a marathon run and percentage type I fibers is 0.93 (Komi et al., 1981). In our laboratory, there was no significant relationship between percentage of type I fibers and MAC/kg in 98 sedentary subjects (Lortie et al., unpublished data). However, individuals with a high proportion of type I fibers (> 60%) achieved a significantly greater work output in 90 minutes than those with a low proportion of type I fibers (> 25%).

Significant associations have been found between muscle oxidative enzyme activities (Booth & Narahara, 1974; Ivy et al., 1980; Rusko, Havu, & Karvinen, 1978) and muscle capillary supply (Ingjer, 1978), and MAP. Unpublished data from our laboratory (Hamel et al., 1983) showed that MAC/kg was significantly correlated with vastus lateralis malate dehydrogenase (MDH), 3-hydroxy-acyl- CoA dehydrogenase (HADH), and 2-oxoglutarate dehydrogenase (OGDH) enzyme activities in 75 sedentary subjects ($r \geqslant 0.33$). About 20% of the total variance in MAC/kg was accounted for by the activities of these three enzymes.

Endurance training influences the metabolic capacity of the skeletal muscle. Thus, the capacity to oxidize carbohydrates and free fatty acids increases with endurance training, as do gluconeogenic and the glycogenic potentials (Howald, 1982). Training experiments conducted with human subjects also indicate that a progressive shift of type IIb fibers toward type IIa fibers is possible (Anderson & Henriksson, 1977; Green, Thomson, Daub, Houston, & Ranney, 1979; Schantz, Billeter, Henriksson, & Jansson, 1982). Even though it has been suggested that endurance training can help transform fiber types in the direction of type I, on the basis of longitudinal training studies in humans the evidence is still inconclusive (Howald, 1982).

Table 3 describes some of our results concerning the effects of E (the same 20-week aerobic training program) on fiber types and areas and enzyme activities of the *vastus lateralis* muscle (Lortie, Simoneau, Hamel, Boulay, & Bouchard, unpublished data). The proportion of type I fibers did not change with training, but that of type IIa fibers and related surface area increased significantly. In addition, muscle MDH activity and the ratio of phosphofructokinase (PFK) to OGDH were significantly modified with the endurance training program. Changes in HADH and OGDH were positive but not significant, an observation that may be related to the relatively short duration of the training program as suggested in animal experiments (Benzi, 1981). With the exception of MDH activity (47%), other muscle variables changed on the average from about 15% to 20% as a result of endurance training.

Substrate availability in the muscle, as well as in the blood and liver, to be used by the working skeletal muscle is central to the understanding of variations in MAP and perhaps even more so in MAC. Blood concentrations of free fatty acids and glycerol increase during prolonged submaximal work, suggesting a greater release from adipose tissue depots. We have demonstrated that suprailiac fat cell basal and epinephrine maximally stimulated lipolysis

Table 3. The influence of a 20-week aerobic training program on skeletal muscle characteristics in 24 subjects ascertained as sedentary[a]

Variable	Pretraining M (SD)	Posttraining M (SD)	Individual changes in % M (SD)	Range in %	Correlation of % gain with pretraining value (r)
% type I fiber	37 (13)	37 (11)	10 (52)	−44 to 156	−0.6**
Area type I fiber (μm^2)	3942 (923)	4401 (939)*	16 (27)	−40 to 66	−0.6**
% type IIa fiber	39 (10)	45 (9)*	19 (29)	−38 to 80	−0.8**
Area type IIa fiber (μm^2)	3836 (1253)	4297 (1202)*	17 (28)	−36 to 73	−0.6**
MDH	133 (34)	183 (32)**	47 (45)	−17 to 160	−0.8**
HADH	3.8 (1.2)	4.2 (1.1)	23 (48)	−55 to 117	−0.7**
OGDH	0.78 (0.33)	0.83 (0.33)	22 (64)	−64 to 166	−0.7**
PFK/OGDH	158 (63)	119 (35)**	−15 (38)	−70 to 101	−0.7**

[a]Adapted from Lortie et al. (unpublished data).
[b]All activities expressed in μmol NADH (NADPH)/g wet weight·min^{-1}.
*$p \leq 0.05$; **$p \leq 0.01$.

were enhanced as a result of a 90-minute maximal cycle ride. On the other hand, fat cell lipogenesis from glucose was diminished while the heparin releasable fraction of adipose tissue lipoprotein lipase was increased following a MAC performance (Savard et al., unpublished data).

Moreover, Després et al. (1983) have demonstrated that lean marathon runners were characterized by elevated isolated fat cell lipolysis in the basal state and when maximally stimulated with epinephrine in comparison with sedentary controls. Long-distance runners were also characterized by increased adipose tissue lipid accumulation activities, as revealed by the elevated lipogenesis from glucose and lipoprotein lipase activity in comparison with a group of sedentary controls (Savard et al., unpublished data).

Costill, Fink, Getchell, Ivy, and Witzmann (1979) have observed that endurance runners exhibit a 7-fold increase in the capacity of the gastrocnemius muscle to oxidize ^{14}C-palmityl CoA over sedentary controls. This enhanced capacity to oxidize fat is thought to be a decisive factor favoring the sparing of muscle glycogen during prolonged intensive exercise. The importance of this phenomenon has been emphasized by several investigators, including Pernow and Saltin (1971) who found a decrease of about 45% in a 60-minute endurance performance in subjects who had a reduced muscle glycogen concentration. It should also be noted that endurance trained athletes exhibit a higher muscle glycogen concentration at rest than untrained individuals (Gollnick et al., 1973). All these factors in substrate availability and utilization during prolonged exercise are generally viewed as related to endurance performance. To some extent they also appear to respond to an endurance training stimulus (Gollnick & Saltin, 1982).

Table 4 illustrates the effects of a 20-week aerobic training program on collagenase-isolated fat cell lipolysis (Després et al., 1984b). Cells were obtained from a suprailiac fat biopsy. Individual changes with training reached a mean of 21% for basal lipolysis (p > 0.05) and 77% for epinephrine maximal stimulated lipolysis (p < 0.01). Correlations between changes in lipolytic activities and pretraining scores attained about 25% of common variance. Although results are not shown here, male subjects improved more in terms of lipolytic activities than females (Després et al., 1984b).

In summary, MAP, MAC, and most of their determinants generally improve with aerobic training. There are indications that MAC might be more respon-

Table 4. The influence of a 20-week aerobic training program on isolated fat cell lipolytic activities in 22 subjects ascertained as sedentary

Variable	Pretraining M (SD)	Posttraining M (SD)	Correlation of % gain with pretraining value (r)
Basal lipolysis[a]	0.20 (0.09)	0.21 (0.08)	−0.5*
Epinephrine maximal stimulated lipolysis[a,b]	1.08 (0.49)	1.69 (0.67)**	−0.5*

Note. Data compiled from Després et al., 1984, and Bouchard, 1983.
[a]Lipolysis in μmol glycerol/30min/10^6 cells.
[b]10^{-4} M epinephrine.
*p ≤ .05; **p < .01.

sive than MAP to training in sedentary subjects. From a population point of view, and with individuals exposed to aerobic training of moderate intensity for 3 months and more, MAP/kg should yield an average training increase of about 20% but with considerable variation (i.e., from about 10% to 40%). For MAC/kg, the same E effect will probably reach about 40% of the mean pretraining value, with a range of training gains from about 10% to 60% from study to study.

The G Effects

The genetic effect for a given multifactorial phenotype can be considered as the average contribution of the genes in the population irrespective of environmental factors and lifestyle. In other words, the G effect is a population parameter reflecting largely the extent of the transmissible genetic variance from one generation to the next. It simply expresses the relative contribution of genetic variation to the interindividual differences seen for the phenotype in the population under the average environmental conditions. Therefore, the G effect does not imply that P is inherited to the extent of G in a given individual. The knowledge of G for a given P has little predictive value as to an individual's genetic endowment level.

The estimate of the population parameter G for a given P requires a complex data base and is not a simple task. One would like to rely on data gathered on all types of two- and three-generation relatives including adoption data, twins reared together and apart, and families of MZ twins. Moreover, large sample sizes would be desirable to derive reliable estimates of G and related parameters. Data currently available seldom meet these stringent conditions, and thus any estimates of G for aerobic power and capacity must be viewed with caution at present.

Maximal Aerobic Power and Capacity

The literature dealing with the G effect in aerobic performance was recently reviewed (Bouchard & Malina, 1983a; Bouchard & Lortie, 1984). First, no study of maximal aerobic capacity has been reported thus far. Second, data from studies of estimated or measured maximal aerobic power with spouses, parents and offspring, and brothers and sisters have supported the hypothesis that the genetic effect reaches a maximum of about 30% to 40% of the total phenotypic variance (Lortie et al., 1982; Montoye & Gayle, 1978). Third, twin studies dealing with the power of the system have reported inconsistent results, with estimates of G ranging from as low as zero to more than 90%. Such discrepancies are thought to be associated with the twin sample sizes, uncontrolled age or gender effects, laboratory methods, and differences in mean or variance between twin types (Bouchard & Malina, 1983a).

At this point it may be useful to introduce some data from our laboratory on the G effect for MAP and MAC, and some of their determinants. Table 5 summarizes data from Lortie et al. (1982) and Lesage et al. (in press) for estimated and directly measured $\dot{V}O_2$ max/kg in several kinds of relatives. In general, correlations for measured MAP/kg are lower than those for estimated MAP/kg except for the mother-child correlation. Parent-child coefficients (2 x r) and twin correlations (2 [$r_{MZ} - r_{DZ}$]) suggest that the G effect is probably quite small for MAP/kg ($0.06 \leqslant G \leqslant 0.34$). Sibling data point to a higher G effect, but they are notoriously inflated. The difference between the mother-child ($r = 0.28$) and the father-child ($r = -0.01$) correlations for measured MAP/kg is interesting and may suggest a maternal effect for this phenotype. One can build a theoretical case to account for this effect by mitochondrial DNA which is known to be maternally inherited (Giles, Blanc, Cann, & Wallace, 1980). As this 16-kb base-pairs DNA is known to code for 2 rRNAs, 22 tRNAs, and 13 mRNAs apparently associated with the maintenance and function of the mitochondrion, this observation certainly merits more attention.

Results of some of our published and unpublished data for direct measurements of aerobic performances in male DZ and MZ twins are summarized in Table 6. As previously shown (Klissouras, 1971), the G effect in max HR appears to be quite high. In the present case the broad heritability coefficient reaches about 70%. It must be noted however, that other studies

Table 5. A summary of data reported from our laboratory about the correlation in pairs of relatives for estimated or measured MAP/kg[a]

Types of relatives	MAP/kg estimated N pairs	r	MAP/kg measured N pairs	r
Spouses	119	.33*	20[b]	.20
Parent-child	564	.17*	109[b]	.03
Father-child	307	.17*	60[b]	−.01
Mother-child	257	.17*	49[b]	.28
Biological sibs	223	.33*	47[b]	.19
DZ twins			20	.73*
MZ twins			37	.76*

[a]From Lortie et al. (1982) and Lesage et al. (in press). Residuals of age and sex.
[b]Data obtained in a treadmill test; all others in a cycle ergometer test.
*$p \leqslant 0.01$.

Table 6. Intrapair resemblance for aerobic performance in MZ and DZ male twins[a]

Variable[b]	DZ twins F ratio	r_i	MZ twins F ratio	r_i	H_B[c]
MAP/kg	6.6**	0.74	9.2**	0.80	0.12
Max HR	2.7*	0.45	9.6**	0.81	0.72
VT-1/kg	3.2**	0.52	5.1**	0.67	0.30
VT-2/kg	3.2**	0.52	4.6**	0.64	0.24
MAC/kg	4.4**	0.63	8.4**	0.79	0.32

[a]20 pairs of DZ and 22 pairs of MZ twins; adapted from Lesage et al. (unpublished data).
[b]All subjects were males from 16 to 28 years.
[c]From Falconer (1960).
*$p < 0.05$; **$p \leqslant 0.01$.

Table 7. Resemblance between relatives for submaximal power output (PWC$_{150}$/kg body weight)[a]

Relatives	N cells (N subjects)	PWC$_{150}$/kg weight F ratio	Intraclass correlation
Spouses	276 (552)	1.5*	.19
Adoptive sibships	46 (107)	1.0	.00
Cousin sibships	33 (87)	1.4	.14
Biological sibships	225 (531)	1.8**	.25
DZ sibships	56 (117)	2.8**	.46
MZ sibships	54 (110)	4.0**	.60

[a]Modified from Bouchard et al. (1984); on residuals of age and sex.
*$p < 0.05$; **$p \leqslant 0.01$.

have not found any indication of an inheritance component for max HR (Klissouras, Pirnay, & Petit, 1973; Komi & Karlsson, 1979).

On the other hand, MAP/kg, MAC/kg, VT-1/kg, and VT-2/kg were characterized by much lower genetic effects ($0.12 \leqslant G \leqslant 0.32$) in the present data base. Even though age and gender were controlled in this data set, remember that sample sizes tended to be small.

In a recent paper, PWC$_{150}$ measurements in biological and adopted sibships were considered and some of these data are summarized in Table 7 with new data for spouses (Bouchard et al., 1984). The analysis of variance procedure comparing variances between sibships (or pairs) and within sibships (or pairs) showed that spouses resemble each other ($p < 0.05$) for PWC$_{150}$/kg. Spouse resemblance for such a trait is probably associated with positive assortative mating for the trait (or a relevant covariate) and/or shared lifestyle (e.g., similar exercise habits). In contrast, sibships of adoptees and of first-degree cousins did not exhibit significantly more variance between sibships than within sibships. However, sibships of biological brothers and sisters, DZ twins, and MZ twins were characterized by significant levels of resemblance ($p \leqslant 0.01$) for PWC$_{150}$/kg. DZ and MZ data suggested that submaximal power output measurements were affected by only a low G effect, that is, about 25% to 30% (Table 7).

Determinants of MAP and MAC

Little has been reported on the effect of heredity on heart size and structures. Animal studies suggest that the coronary network of the heart is genetically determined (Grewal & Purushothaman, 1978). Human studies suggest that biological inheritance plays a major role in the vascular wall thickness of the left coronary arteries in infants (Personen, Norio, & Sarna, 1975), in the anatomical pattern of the coronary arteries (Motulsky, 1977), and in the branching pattern of pulmonary arteries (Hislop & Reid, 1973). An earlier study dealing with the volume of the heart per kg of body weight (Klissouras et al., 1973) submitted data which, when reanalyzed (Bouchard & Malina, 1983a), yielded intraclass correlations of 0.42 in 23 pairs of MZ twins and 0.28 in 16 pairs of DZ twins. These results translate into a heritability coefficient between 25 and 30%.

Heart dimensions in several kinds of relatives have also been measured in our laboratory by echographic techniques, and some of these results are summarized in Table 8. All measurements were made in the diastolic phase of the heart cycle. For two of the variables, left ventricle internal diameter and left ventricular volume per m^2 of body surface area (BSA), the correlation pattern was quite clear. Individuals living together but who shared no genes by immediate descent did not covary significantly for these measurements, the correlations being generally near zero. In contrast, for pairs of individuals sharing about 50% of their genes and cohabitational conditions (i.e., parent-child, biological brothers and sisters, and DZ twins), the coefficients were significant and generally above 0.2. For these two measurements, the highest correlations occurred in MZ pairs ($r \geq 0.59$). Such a pattern was not clearly evident in the three other cardiac dimensions of the study. The data suggest that the latter heart measurements are either meaningfully influenced by conditions prevailing in the living environment or affected by a concomitant variable that has escaped our attention. From this data set, however, it can be estimated that left ventricular volume, after control over BSA, is characterized by a G effect possibly ranging from 30% to 70%.

Medical geneticists have demonstrated that muscular properties can be the object of inherited influences as indicated in the catalog of known gene-associated muscle defects (McKusick, 1982). Structural genes for contractile proteins of skeletal muscle are numerous. Variations in the expression of these genes appear responsible for the coexistence of mixed forms of the contractile proteins in a given fiber type. This was suggested recently for myosin and troponin in human skeletal muscle (Billeter, Heizmann, Howald, & Jenny, 1981; Billeter et al., 1980). Although gene expression, and perhaps even genetic variation in the structural genes involved, seem to be of primary significance in muscle fiber distribution within a muscle, available studies have not dealt with these issues in humans. Most of the studies reported to date are based on the classical twin method and the results are perhaps affected by some of the problems identified previously (Bouchard & Malina, 1983a). Komi et al. (1977) noted that MZ twins (15 pairs) were essentially similar in fiber type distribution, while DZ twins (16 pairs) were quite variable. From their data, heritability coefficients for the percentage of type I fibers in the vastus lateralis reached 0.93 or higher.

Table 9 summarized some data on covariation between biological sibs for skeletal muscle characteristics (see Lortie et al., 1985). Intraclass coefficients in 34 pairs of MZ twins of both sexes reached about 0.55 for the percentage of types I and IIb fibers in the vastus lateralis muscle. The coefficients were statistically significant but far from the level of resemblance reported earlier for MZ twins by Komi et al. (1977). The same coefficients in 32 pairs of brothers reached about 0.3. In contrast, resemblance in these two kinds of biological siblings for the proportion of type IIa fibers was nonexistent.

In ultrastructural analyses of skeletal muscle of 11 MZ and 6 DZ twin pairs, Howald (1976) reported no gene-associated variation in the mitochondrial density, the ratio of mitochondrial volume to myofibril volume, and the internal and external surface densities of mitochondria. Furthermore, activities of hexokinase (HK) and succinate dehydrogenase (SDH) were identical in both sets of twins, while muscle glyceraldehyde-3-phosphate dehydrogenase and HADH

Table 8. Interclass correlations in pairs of cultural and/or biological relatives for selected echographic measurements of the heart dimensions[a]

Variable[b]	Spouses (N ≥163)	Unrelated sibs (N ≥74)	Parent-adopted child (N ≥190)	Parent-child (N ≥647)	Biological sibs (N ≥220)	DZ twins (N =41)	MZ twins (N =38)
Left ventricle internal diameter (mm)	−.11	.14	.09	.20**	.29**	.41**	.61**
Left ventricle posterior wall thickness (mm)	.22**	.06	.20**	.25**	.21**	.57**	.42**
Interventricular septum thickness (mm)	.22**	.44**	.38**	.34**	.32**	.34**	.52**
Left ventricular volume/ BSA (ml/m²)	−.09	.07	.05	.15**	.22**	.25	.59**
Left ventricular mass/ BSA (g/m²)	.17*	.42**	.25**	.29**	.37**	.28	.56**

[a]Data from Diano et al. (see Bouchard & Lortie, 1984) and unpublished study from our laboratory; N = number of pairs.
[b]On scores adjusted by generation for the effects of age and sex.
*p ≤ 0.05; **p ≤ 0.01.

Table 9. Intraclass coefficients in pairs of biological sibs for fiber type distribution and selected enzyme activities from the vastus lateralis muscle[a]

Variable	Brother-brother (N = 32 pairs)[b] F ratio	r intra	MZ twins (N = 35 pairs)[c] F ratio	r intra
% type I	2.2*	0.33	3.4**	0.55
% type IIa	0.9	−0.03	1.4	0.18
% type IIb	1.8	0.26	3.6**	0.56
HK	0.6	−0.22	2.4**	0.41
PFK	1.8	0.27	3.4**	0.55
MDH	1.4	0.15	3.8**	0.58
OGDH	1.2	0.09	3.2**	0.53
PFK/OGDH	2.1*	0.34	1.9*	0.00

[a]Adapted from Lortie et al. (this volume).
[b]Scores adjusted for age differences.
[c]Scores adjusted for age and sex effects.
*$p \leqslant 0.05$; **$p \leqslant 0.01$.

were significantly more variable among DZ twins than among MZ twins. With one member of a MZ twin pair (n = 7 pairs) participating in a 23-week endurance training program and the other maintaining his usual activities, Howald noted significant increases in selected extra- and intramitochondrial enzymes. For example, SDH and HK activities increased by 28% and 17%, respectively, with training in one member of the MZ pair compared to the other. Moreover, in the study reported by Komi et al. (1977), there was no evidence of a significant G effect in activities of several skeletal muscle enzymes from DZ and MZ twin comparisons.

In Lortie et al.'s study (1985), coefficients for muscle enzyme activities were not characterized by a systematic pattern that could reflect a strong G effect (see Table 9). MZ twin coefficients were generally moderate but statistically significant. Biological brother resemblance was lower and significant only for the PFK/OGDH ratio. Data for PFK, MDH, and OGDH are compatible with a significant G effect, but they are not sufficient by themselves to demonstrate it clearly. The fact that MZ correlations for these enzyme activities reach only about 0.55 suggests that nongenetic factors are also at work. Additional data on different kinds of relatives are needed before more definite conclusions can be drawn on the size of the G effect in skeletal muscle aerobic and anaerobic-to-aerobic characteristics.

There have been few reports on substrate availability and utilization in the context of genetic variation in healthy, exercising humans. However, data accumulated on inborn errors of metabolism indicate that genetic variation can affect the flow of substrates in various tissues. Thus, inherited disorders of glycogen metabolism (Mahler, 1976) as well as of lipid metabolism in skeletal muscle (Angelini, 1976) have been described. Most of these inborn disorders are present in homozygotes for a recessive gene which translates into a dramatic decrease, if not a total loss, of enzyme activity. Animal studies have also shown that substrate concentration is sometimes determined by the genes. For instance,

Dalrymple, Kastenschmidt, and Cassens (1973) have observed breed differences in pigs for the glycogen concentration of the white semimembranous muscle, but not the red trapezius muscle.

Preliminary evidence for significant MZ twin resemblance (Després & Bouchard, 1984) and lack of resemblance in DZ twins (Bouchard, Després, Savard, & Leblanc, unpublished data) has been reported for isolated fat cell basal and maximal epinephrine stimulated lipolysis in tissue obtained from the suprailiac region. These data are summarized in Table 10. They did not provide a clear picture about the G effect for the fat mobilization process but suggested that a common environment may be one of the decisive factors. On the other hand, glucose conversion into triglycerides in collagenase isolated fat cells was characterized by almost zero intraclass coefficients in basal conditions, but by a significant MZ within-pair resemblance when maximally stimulated with insulin (r = 0.70). In the resting fasting state, the same trend was observed for the lipoprotein lipase activity of adipose tissue (r = 0.78 for MZ twins). After exposure to a 90-minute maximal cycle exercise, DZ twin resemblance did not change for any of the adipose tissue metabolic parameters, but MZ twin resemblance decreased for basal and for stimulated lipolysis and increased for glucose conversion into triglycerides under basal conditions (Savard et al., unpublished data).

In summary, the G effect, which includes the additive genetic effects plus other effects as well, tends to be rather moderate and sometimes quite small for maximal aerobic power, maximal aerobic capacity, and some of their deter-

Table 10. Intraclass coefficient in male DZ and MZ twins for fat cell metabolism at rest and after prolonged (90-min) exercise[a]

Variable	MZ twins $r$intra (N ≥ 12 pairs)	DZ twins $r$intra (N ≥ 11 pairs)
Resting values		
BL	.68*	.09
EML	.59*	.07
BLG	.08	.17
ILG	.70*	.30
LPL	.78*	.25
After prolonged exercise		
BL	−.20	.18
EML	.18	.07
BLG	.97*	.25
ILG	.90*	.33
LPL	.71*	.33

[a]From Després et al. and Savard et al. (unpublished data). BL = basal lipolysis; EML = epinephrine (10^{-4} M) stimulated lipolysis; BLG = basal glucose conversion into triglycerides; ILG = insulin (9 μU/ml) stimulated glucose conversion into triglycerides; LPL = adipose tissue heparin releasable lipoprotein lipase activity. BL and EML in μmol glycerol/30 min/10^6 cells; BLG and ILG in nmol glucose/h/10^6 cells. LPL in μmol FFA/h/10^6 cells.
*p ≤ 0.01.

minants. It should also be remembered that in several cases the data base remains quite tenuous to draw conclusions about the population parameter G.

The GxE Effect

At this point, it is appropriate to turn to the GxE effect. Figure 1 describes the response to training for MAP and MAC as commonly understood by sport scientists, coaches, and fitness leaders. In this model there are differences in MAP/kg or MAC/kg associated with a G effect, but a G effect that is indepen-

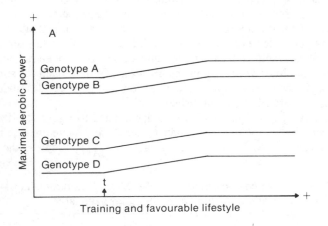

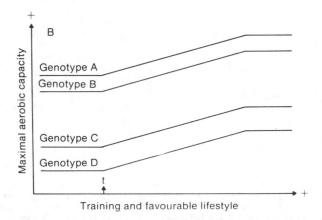

Figure 1. Models of sources of variation in aerobic performances commonly accepted in the sports sciences literature. In this case P = G + E, and GxE = 0. The E effect is larger in maximal aerobic capacity than in maximal aerobic power; t stands for the threshold of training.

dent of training. In addition there is an E effect, which is the same for all genotypes when training goes beyond the threshold point.

Also in this model, a given amount of E has the same effect on MAP (panel A), regardless of G, that is, there is no GxE component and all individuals exhibit the same sensitivity to E. This critical assumption is commonly accepted in the sport sciences literature but cannot be justified based on recent findings. The model also assumes that all G have the same maximal trainability of MAP and that all will reach a plateau in response to training at the same E level. Again, these are among the "hidden" assumptions of the sport sciences literature that must be challenged in view of recent laboratory findings.

Figure 1 also suggests that the same basic model applies to MAC or human variation in endurance performance (panel B). However, there are two important differences when comparing MAC with MAP (i.e., panels B and A). First, the effects of E on MAC are felt for a longer period of E than for MAP. In other words, the plateau phase for MAC comes at a later amount of E. Second, MAC is more trainable than MAP. Nevertheless, the assumptions underlying this model are weakened by the fact that all G are made to behave similarly as it is still commonly assumed in the physical activity sciences.

Let us consider some recent findings concerning human variation in the adaptive response of aerobic performance to E. For instance, in the study of Lortie et al. (1984), 24 sedentary subjects (based on interview and questionnaire) trained for 20 weeks under standardized conditions. Mean improvement in maximal aerobic power was 33%, with gains ranging from 5% to 88%. In the case of MAC/kg, mean training improvement was 51%, with a range of 16% to 97%. In other words, there were high responders and low responders to an endurance training program designed to improve the power and capacity of the aerobic work metabolism. Similar findings were reported for epinephrine maximal stimulated lipolysis (EML) on collagenase isolated fat cells (Després et al., 1984a; Bouchard, 1983). Thus, after 20 weeks of aerobic training some individuals did not show any changes in EML, whereas several increased their EML activity by more than 100% and one subject by more than 300%. In other words, regardless of the P of the aerobic metabolism, considerable variation in adaptation to training were apparent.

The question then becomes, What is responsible for differences in adaptation to aerobic training? Figure 2 describes a model of factors potentially related to human variation in sensitivity to aerobic training. These factors include age and sex of subjects, previous training experience, current phenotype level, and heredity, but they have been reordered from a previous version of the model (Bouchard & Malina, 1983; Bouchard & Lortie, 1984) to clarify the contribution of the inherited component.

Age of subjects is a readily controlled factor in laboratory experiments, and the results from our group (Bouchard, Carrier, Boulay, Thibault-Poirier, & Dulac, 1975; Bouchard, 1983; Després et al., 1984; Lortie et al., 1984; Prud'homme, Bouchard, Leblanc, Landry, & Fontaine, 1984b) as well as from other laboratories clearly demonstrate that the response to aerobic training remains highly variable within a relatively narrow age range. The factor labeled as "previous experiences" has received little attention from the sport sciences community. The question is, would individuals with a past experience of aerobic training respond differently (presumably better) from those who have had no

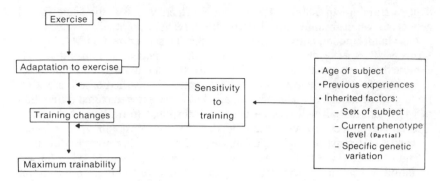

Figure 2. A model of factors associated with human variation in the sensitivity to exercise-training (Modified after Bouchard & Malina, 1983).

such training? Unfortunately, this question cannot be answered adequately at this time. One way to avoid the possible confounding effect of this factor is to use experimental subjects ascertained as sedentary with no history of regular participation in sports or other physical activities. This is what we have elected to do in experiments designed to quantify average E effects and individual differences in response to training. Under these conditions, that is, a narrow age range and no prior training experience, the evidence indicated considerable individual differences in the training changes in MAP/kg and MAC/kg within each sex (Lortie et al., 1984).

The remaining three factors are associated in some ways with the genotype of the individual. In their aerobic training experiment, Lortie et al. (1984) have concluded that MAP had the same trainability in both males and females. On the other hand, they found that not only was MAC about 60% more trainable than MAP on the average, but that males enjoyed a better trainability of MAC than females under similar laboratory conditions. In other words, sex of subjects may account for some of the variations observed in the response of MAC to training.

One of the most difficult factors to interpret is the current phenotype level, that is, the pretraining level in the present context and its relationship with the response to training. Several studies have shown that trainability of maximal aerobic power and other aerobic output criteria is negatively related to the pretraining level of the attribute (Bouchard et al., 1975; Bouchard, Boulay, Thibault, Carrier, & Dulac, 1980; Pollock, 1973; Saltin, 1972). In a review of data published from 50 training experiments, the correlation between training-induced changes in MAP/kg and the pretraining level of the attribute was significant, $r = -0.5$ (Bouchard et al., 1975). When statistical controls were applied to frequency, intensity, and duration of training sessions, and to the number of weeks of training, the correlate remained significant, $r = -0.4$.

More recently, the correlation between training gains in MAP/kg and initial $\dot{V}O_2$ max per kg body weight in sedentary subjects was about -0.6, while that between training gains in MAC/kg and initial level was about -0.5 (Lortie et al., 1984; see Table 1). These results imply that the initial phenotype level accounts for a significant fraction of the variance observed in response to aerobic training. But the current phenotype level, when dealing with subjects confidently ascertained as sedentary, is partly a reflection of the G effect. The variance in the training response associated with the pretraining P can thus under these circumstances be conceived as a component of the GxE effect.

Finally, the fifth factor is the genotype-training interaction component associated with genetic variation(s) unknown at this time but independent of pretraining differences in P. This is represented by the term GxE and it implies that the sensitivity of MAP or MAC to aerobic training depends to some extent on the genotype of the individuals. From early data, doubts were expressed regarding the presence of an interaction effect in the training of MAP/kg (Klissouras, 1976; Weber, Kartodihardjo, & Klissouras, 1976). But there is now sufficient evidence to support the hypothesis that GxE is present for several indicators of aerobic work metabolism. In addition, when GxE is investigated with an appropriate design it turns out to be a major component of the variance in the training response. Some of these studies are considered subsequently.

Ten pairs of MZ twins were subjected to a 20-week endurance training program, initially meeting four and later five times per week. Training sessions lasted 40 and then 45 minutes, with training intensity averaging 80% of the maximal heart rate reserve (starting at 60% and increasing to 85%). Under this program, maximal aerobic power improved by 14% while the ventilatory thresholds increased by 17%. However, there were considerable interindividual differences in training gains as illustrated by the range of 0 to 41% for $\dot{V}O_2$ max per kg body weight. But differences in the response to training were not distributed randomly among the twin pairs. Thus, intraclass correlations computed with the amount of gain as a percentage of pretest MAP/kg reached 0.82, indicating that members of the same twin pair yielded a fairly similar response to training; that is, 82% of the variance in the training response was genotype-dependent. These results suggest that the sensitivity of maximal aerobic power to training is largely genotype-dependent (Bouchard, 1983; Prud'homme et al., 1984b). Figure 3 illustrates these results for training gains in percentage of $\dot{V}O_2$ max/kg ($r_{intra} = 0.82$). On the other hand, as shown in Table 11, this effect appeared to be lower for submaximal indicators of VT-1 and VT-2, the intraclass coefficients being 0.54 ($p < 0.05$) and 0.33, respectively.

Recent data also support the contention that adaptive responses in substrate availability and utilization to endurance training are largely genotype-dependent. For instance, Després et al. (1984a) reported on the effects of 20 weeks of endurance training in 8 pairs of MZ twins. Basal and epinephrine submaximal and maximal stimulated lipolysis were determined on isolated fat cells before and after training. A genetically determined response to endurance training was not found for basal lipolysis. The magnitude of changes in stimulated lipolysis was very heterogeneous, with a range from −43 to +317%. The within-MZ-pair resemblance in the sensitivity of maximal lipolysis, however, was very high, with an intraclass correlation of 0.90 (Bouchard, 1983; Després et al., 1984a). Therefore it was concluded from these experiments that the sensitivity of stimulated lipolysis to endurance training is genotype-dependent (Table 12).

In a more recent experiment from our laboratory and reported elsewhere in this volume, Boulay, Lortie, Simoneau, and Bouchard committed 14 pairs

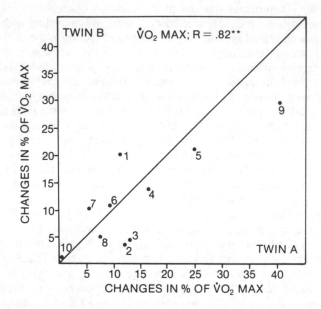

Figure 3. Intrapair resemblance (intraclass coefficient) in 10 pairs of MZ twins for training changes after 20 weeks in percent of $\dot{V}O_2$ max/kg. (**for $p < 0.01$.)

of MZ twins to a 15-week intensive anaerobic training program. MAP/kg and MAC/kg improved by 22% and 17%, respectively. Interindividual differences were again considerable, with changes in $\dot{V}O_2$ max per kg body weight ranging from about 0 to 65% and improvements in performance on the 90-minute test ranging from about 0 to 55%. The responses of these criteria of aerobic performance to an anaerobic training program, however, were significantly

Table 11. The effect of a 20-week aerobic training program on measurements of aerobic performance in 10 pairs of MZ twins and intraclass correlation for twin resemblance in the magnitude of training response[a]

Variable	Pretraining values M	(SD)	Posttraining values M	(SD)	Intraclass twin resemblance in response (% gain)
Max (ml O_2/kg·min⁻¹)	44.2	(6.0)	49.7**	(6.0)	0.82**
VT-1 (ml O_2/kg·min⁻¹)	25.0	(3.3)	29.8**	(4.1)	0.54*
VT-2 (ml O_2/kg·min⁻¹)	36.0	(4.9)	41.5**	(5.6)	0.33

[a]Adapted from Prud'homme et al. (1984b).
*$p < 0.05$; **$p \leqslant 0.01$.

Table 12. The effect of a 20-week aerobic training program on adipocyte lipolytic activities in 8 pairs of MZ twins and intraclass correlation for twin resemblance in the magnitude of training response[a]

Variable	Pretraining values M	(SD)	Posttraining values M	(SD)	Intraclass twin resemblance in response (% gain)
Basal lipolysis [b]	0.30	(0.08)	0.46*	(0.11)	0.17
Epinephrine (10^{-5}M) stimulated lipolysis	0.52	(0.19)	0.88*	(0.26)	0.84*
Epinephrine (10^{-4}M) stimulated lipolysis	1.38	(0.61)	2.15*	(0.76)	0.90*

[a]From Després et al. (1984).
[b]Lipolysis expressed in μmol glycerol/30 min/10^6 cells.
*$p \leqslant 0.01$.

similar in members of the same twin pair ($0.44 \leqslant r_{intra} \leqslant 0.69$), suggesting again a biologically and quantitatively meaningful genotype dependency.

The presence of a meaningful GxE effect implies that there are high (HR) and low responders (LR) to aerobic training. As predicted, such differences in the response of MAP/kg to training were found in association with certain genotypes. This is illustrated by the cases presented in Table 13. In addition, early responders and late responders were also identified in the course of a 20-week aerobic training program, both types being found in the HR and in the LR categories (Bouchard, 1983). Table 14 describes some of these cases whose differences in the rate of the response of MAP/kg to training were apparently associated with the genotype. Additional and more extensive data are needed to identify with more certainty the various patterns associated with this GxE effect. Data on the same phenomenon for MAC/kg are not yet available but soon will be.

Figure 4 summarizes in a schematic way the general trends uncovered thus far concerning the response of MAP/kg to aerobic training. In this case, varia-

Table 13. Illustration of characteristically high and low responses to a 20-week aerobic training program in female MZ pairs[a]

MZ pair	Max ml O_2/kg·min^{-1} pretraining	Gain in ml O_2 after 20 weeks
MZ 10 (FED)	37	11
MZ 10 (FID)	34	14
MZ 15 (CB)	39	0
MZ 15 (HB)	41	1

[a]From Bouchard (1983).

Table 14. Illustration of characteristically early and late responses to a 20-week aerobic training program in male MZ pairs[a]

MZ pair	Max ml O_2/kg·min⁻¹ pretraining	Gain in ml O_2/kg·min⁻¹ After 7 weeks	After 20 weeks
MZ 19 (CG)	41	8	5
MZ 19 (SG)	44	6	6
MZ 07 (SAL)	52	0	6
MZ 07 (SYL)	53	0	3

[a]Adapted from Bouchard (1983) and unpublished data.

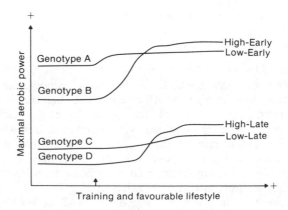

Figure 4. Schematic summary of trends in the response of maximal aerobic power to training. In this case, P = G + E + GxE. Arrow represents the initiation of aerobic training (adapted after Bouchard & Lortie, 1984).

tions in P with training are accounted for by G, E, and GxE effects. This situation may turn out to be rather realistic as the influence of the pretraining level is partially accounted for by GxE; sex of subject was not significant in our experiments with MAP/kg, and age and prior experience were controlled for by design.

In Summary: Human Variation in MAP and MAC

Even though the research is not yet completed, how can the data best be summarized at this stage? From the findings reported here, it seems reasonable to conclude that the various effects associated with variation in P are different from a quantitative point of view in MAP and in MAC. Table 15 describes the parameter estimates for MAP/kg and MAC/kg as we believe they are at present from our research on sedentary individuals of both sexes submitted to training.

When considering MAP/kg, most of the population variance is genotype-dependent, provided age does not contribute significantly to the data. As for the influence of prior experience in training for MAP/kg, we know little about its contribution but presumably it is a minor component. In any case it has been set to zero. Under these conditions, the mean E effects reaches about 20%, the G effect accounts for about 30% of the variance, and the total GxE effect reaches about 50%. On the other hand, the pretraining effect, which is also largely G dependent if not entirely so in sedentary subjects, generally accounts for about 10% of the variance or about 20% of GxE.

The parameters are slightly different for MAC/kg, as shown in Table 15. The E effect reaches about 40% of the variance while the G effect is slightly lower than in MAP/kg. All three interaction components together account for about 40% of the population variance, the main GxE effect reaching about 25% of the variance. Even though there is no gender effect in MAP/kg, one can recognize such an effect in MAC/kg, at least on the basis of the data currently available.

Similar estimates can also be derived for some of the determinants of aerobic performance considered earlier in this review (i.e., heart dimensions, muscle fiber types and enzyme activities, substrate availability). However, data are generally less abundant than for MAP/kg and MAC/kg such that the exercise would be premature. Nonetheless, it is useful to recognize at this stage that quantitatively important GxE effects have been observed in our laboratory for selected skeletal muscle enzyme activities and fat cell stimulated lipolysis. This area certainly deserves more experimental work.

A major problem we have not yet dealt with adequately concerns the identification of individuals who are high (HR) or low responders (LR) to aerobic training. There are at present no genetic markers that one could use to type an individual for sensitivity to training. In other words, sensitivity to aerobic

Table 15. Tentative parameter estimates of causal sources of variation in maximal aerobic power (ml O_2/kg•min^{-1}) (MAP/kg) and maximal aerobic capacity (kJ/kg) (MAC/kg) in sedentary individuals submitted to training[a]

Parameter	MAP/kg (%)	MAC/kg (%)
G effect	30	20
E effect	20	40
G x E effect:		
main G x E effect	40	25
pretraining effect	10	5
sex effect	0	10

[a]Most probable value in a population of sedentary individuals submitted to training, but in our judgment, ranges of values remain possible at this stage, particularly when dealing with small sample sizes and/or individuals of mixed training status.

Table 16. An estimation of the frequency of low responders and high responders to training of aerobic performances[a]

Variable	N cases[b]	% gain with training (Mean)	Range of % gain with training	Low responders: gained less than 5% (N)	8% (N)	High responders: gained more than 40% (N)	60% (N)
MAP/kg	77	23	~ 0 to 88%	3	9	8	2
MAC/kg	63	34	~ 0 to 96%	3	6	19	7

[a]From Bouchard et al. (unpublished data).
[b]Includes both male and female subjects, but only one member of each twin pair.

training of MAP and MAC is still unpredictable. It can only be ascertained from past training experiences or as training progresses. Moreover, the question of the frequency of HR and LR phenotypes has not been addressed. On the basis of several studies of sedentary individuals in our laboratory (Lortie et al., 1984; Prud'homme et al., 1984; Simoneau et al., unpublished data), a reasonable number of subjects have been trained for periods of 15 to 20 weeks. Table 16 summarizes several simple statistics allowing a first estimate of the frequency of LR and HR phenotypes. In the case of MAP/kg, only 3 subjects (or 4%) gained less than 5% with training. Under the same stringent definition of LR, 3 subjects were also in the same category (or 5% of the population) for MAC/kg.

As for the HR phenotype, 2 subjects (3%) gained more than 60% in MAP/kg, while 7 (11%) improved MAC/kg by as much. In other words, 5% to 10% of the population may qualify as HR subjects for various genetic factors. The frequency of LR subjects may be slightly lower (5% or less), but this may only be a problem of definition. The important point to note is that as the HR or the LR phenotypes are clearly associated with inherited factors, they probably belong to these categories of responsiveness to training for a variety of genetic variations and not for a unique genetic variant. In other words, HR and LR phenotypes will most likely turn out to heterogeneous, which increases the complexity of studying them in the laboratory and detecting them in the field.

Conclusion

The attributes of maximal aerobic power and maximal aerobic capacity are highly complex and multifactorial phenotypes. MAP and MAC are trainable properties, particularly in the asymptomatic human organism but undoubtedly in individuals handicapped in one way or another as well. The average training effect is not sufficient by itself, however, to account for the outstanding performances in MAP or MAC of the elite athletes. Favorable genetic variations have to be present so that the performer adapts well beyond the expected training response. The elite performer in aerobic performance has to be well endowed originally (G effect) and has to be a high responder from the genetic point of view (GxE effect and related components); that is, he or she must be an HR subject as well.

Finally, as stated in a recent publication (Bouchard & Lortie, 1984), sport scientists and geneticists should consider investigating in more detail the GxE component in terms of its implications for skeletal muscle, the heart and circulation, substrate utilization, endocrine adaptation, and others. Moreover, the time has arrived to launch major studies into the genetic structure, regulatory elements, and the polymorphism of proteins involved in performance. Fascinating developments in the understanding of the adaptive capacity of the human organism and its genotype dependency are to be expected soon.

References

Andersen, P., & Henriksson, J. (1977). Training induced changes in the subgroups of human type II skeletal muscle fibers. *Acta Physiologica Scandinavica*, **99**, 123-125.

Angelini, C. (1976). Lipid storage myopathies. A review of metabolic defect and of treatment. *Journal of Neurology*, **214**, 1-11.

Benzi, G. (1981). Endurance training and enzymatic activities in skeletal muscle. In P.E. di Prampero & J.R. Poortmans (Eds.), *Physiological chemistry of exercise and training* (pp. 165-174). Basel, Switzerland: Karger.

Billeter, R., Heizmann, C.W., Howald, H., & Jenny, E. (1981). Analysis of myosin light and heavy chain types in single human skeletal muscle fibers. *European Journal of Biochemistry*, **1161**, 389-395.

Billeter, R., Weber, H., Lutz, H., Howald, H., Eppenberger, H.M., & Jenny, E. (1980). Myosin types in human skeletal muscle fibers. *Histochemistry*, **65**, 249-259.

Blomqvist, C.G., & Saltin, B. (1983). Cardiovascular adaptations to physical training. *Annual Review of Physiology*, **45**, 169-189.

Booth, F.W., & Narahara, K.A. (1974). Vastus lateralis cytochrome oxidase activity and its relationship to maximal oxygen consumption in man. *Pflügers Archiv*, **349**, 319-324.

Bouchard, C. (1983). Human adaptability may have a genetic basis. In F. Landry (Ed.), *Health risk estimation, risk reduction and health promotion. Proceedings of the 18th Annual Meeting of the Society of Prospective Medicine* (pp. 463-476). Ottawa: Canadian Public Health Association.

Bouchard, C., Boulay, M.R., Thibault, M.C., Carrier, R., & Dulac, S. (1980). Training of submaximal working capacity: Frequency, intensity, duration and their interactions. *Journal of Sports Medicine and Physical Fitness*, **20**, 29-40.

Bouchard, C., Carrier, R., Boulay, M.R., Thibault-Poirier, M.C., & Dulac, S. (1975). *Le développement du Système de Transport de l'Oxygène chez les Jeunes Adultes*. Québec: Editions du Pélican.

Bouchard, C., & Lortie, G. (1984). Heredity and endurance training. *Sports Medicine*, **1**, 38-64.

Bouchard, C., Lortie, G., Simoneau, J.A., Leblanc, C., Thériault, G., & Tremblay, A. (1984). Submaximal power output in adopted and biological siblings. *Annals of Human Biology*, **11**, 303-309.

Bouchard, C., & Malina, R.M. (1983). Genetics for the sport scientist: Selected methodological considerations. *Exercise and Sport Sciences Reviews*, **11**, 275-305.

Bouchard, C., & Malina, R.M. (1983a). Genetics of physiological fitness and motor performance. *Exercise and Sport Sciences Reviews*, **11**, 306-339.

Bouchard, C., & Malina, R.M. (1984). Genetics and Olympic athletes: A discussion of methods and issues. In J.E.L. Carter (Ed.), *Kinanthropometry of Olympic athletes* (pp. 28-38). Basel, Switzerland: Karger.

Boulay, M.R., Hamel, P., Simoneau, J.A., Lortie, G., Prud'homme, D., & Bouchard, C. (1984). A test of aerobic capacity: Description and reliability. *Canadian Journal of Applied Sport Sciences, 9,* 122-126.

Carter, J.E.L. (1978). Prediction of outstanding athletic ability: The structural perspective. In F. Landry & W.A.R. Orban (Eds.), *Exercise physiology* (pp. 29-42). Miami: Symposia Specialists.

Carter, J.E.L. (1982). Body composition of Montreal Olympic athletes. In J.E.L. Carter (Ed.), *Physical structure of Olympic athletes* (pp. 107-116). Basel: Karger.

Costill, D.L., Fink, W.J., Getchell, L.H., Ivy, J.L., & Witzmann, F.A. (1979). Lipid metabolism in skeletal muscle of endurance-trained males and females. *Journal of Applied Physiology, 47,* 787-791.

Costill, D.L., Fink, W.J., & Pollock, M.L. (1976). Muscle fiber composition and enzyme activities of elite distance runners. *Medicine and Science in Sports, 8,* 96-100.

Dalrymple, R.H., Kastenschmidt, L.L., & Cassens, R.G. (1973). Glycogen and phosphorylase in developing red and white muscle. *Growth, 37,* 19-34.

Després, J.P., & Bouchard, C. (1984). Monozygotic twin resemblance in fatness and fat cell lipolysis. *Acta Geneticae Medicae et Gemellologiae, 33,* 475-480.

Després, J.P., Bouchard, C., Savard, R., Prud'homme, D., Bukowiecki, L., & Thériault, G. (1984a). Adaptive changes to training in adipose tissue lipolysis are genotype dependent. *International Journal of Obesity, 8,* 87-95.

Després, J.P., Bouchard, C., Savard, R., Tremblay, A., Marcotte, M., & Thériault, G. (1984b). The effect of a 20-week endurance training program on adipose-tissue morphology and lipolysis in men and women. *Metabolism, 33,* 235-239.

Després, J.P., Savard, R., Tremblay, A., & Bouchard, C. (1983). Adipocyte diameter and lipolytic activity in marathon runners: Relationship with body fatness. *European Journal of Applied Physiology, 51,* 223-230.

Després, J.P., Bouchard, C., Bukowiecki, L., Savard, R., & Lupien, J. (1983a). Morphology and metabolism of human fat cells: A reliability study. *International Journal of Obesity, 7,* 231-240.

Falconer, D.S. (1960). *Introduction to quantitative genetics.* New York: Ronald Press.

Foster, C., Costill, D.L., Daniels, J.T., & Fink, W.J. (1978). Skeletal muscle enzyme activity, fiber composition and $\dot{V}O_2$ max in relation to distance running performance. *European Journal of Applied Physiology, 39,* 73-80.

Giles, R.E., Blanc, H., Cann, H.M., & Wallace, D.C. (1980). Maternal inheritance of human mitochondrial DNA. *Proceedings of the National Academy of Sciences, U.S.A., 77,* 6715-6719.

Gollnick, P.D., Armstrong, R.B., Saubert, C.W., Piehl, K., & Saltin, B. (1972). Enzyme activity and fiber composition in skeletal muscle of untrained and trained men. *Journal of Applied Physiology, 33,* 312-319.

Gollnick, P.D., Armstrong, R.B., Saubert, C.W., Sembrovich, W.L., Shepherd, R.E., & Saltin, B. (1973). Glycogen depletion patterns in human skeletal muscle fibers during prolonged work. *Pflügers Archiv, 344,* 1-12.

Gollnick, P.D., & Saltin, B. (1982). Significance of skeletal muscle oxidative enzyme enhancement with endurance training. *Clinical Physiology, 2,* 1-12.

Green, H.J., Thomson, J.A., Daub, W.D., Houston, M.E., & Ranney, D.A. (1979). Fiber composition, fiber size and enzyme activities in vastus lateralis of elite athletes involved in high intensity exercise. *European Journal of Applied Physiology, 41,* 109-117.

Grewal, M.S., & Purushothaman, S.C. (1978). Genetic factors in coronary heart disease: Anatomical evidence from mice. *Lancet*, **1**, 603.

Hamel, P., Lortie, G., Simoneau, J.A., Lesage, R., Boulay, M.R., & Bouchard, C. (1983). Relationship between human muscle characteristics and maximal aerobic capacity. *Canadian Journal of Applied Sport Sciences*, **8**, 221.

Hickson, R.C., Bomze, H.A., & Holloszy, J.O. (1977). Linear increase in aerobic power induced by a strenuous program of endurance exercise. *Journal of Applied Physiology*, **42**, 372-376.

Hislop, A., & Reid, L. (1973). The similarity of the pulmonary artery branching system in siblings. *Forensic Science International*, **2**, 37-52.

Howald, H. (1976). Ultrastructure and biochemical function of skeletal muscle in twins. *Annals of Human Biology*, **3**, 455-462.

Howald, H. (1982). Training-induced morphological and functional changes in skeletal muscle. *International Journal of Sports Medicine*, **3**, 1-12.

Ingjer, F. (1978). Maximal aerobic power related to the capillary supply of the quadriceps femoris muscle in man. *Acta Physiologica Scandinavica*, **104**, 238-240.

Ivy, J.L., Withers, R.T., Van Handel, P.J., Elger, D.H., & Costill, D.L. (1980). Muscle respiratory capacity and fiber type as determinants of the lactate threshold. *Journal of Applied Physiology*, **48**, 523-527.

Keul, J., Dickhuth, H.H., Lehmann, M., & Staiger, J. (1982). The athlete's heart—haemodynamics and structure. *International Journal of Sports Medicine*, **3**, 33-43.

Klissouras, V. (1971). Heritability of adaptive variation. *Journal of Applied Physiology*, **31**, 338-344.

Klissouras, V. (1976). Prediction of athletic performance: Genetic considerations. *Canadian Journal of Applied Sport Sciences*, **1**, 195-200.

Klissouras, V., Pirnay, F., & Petit, J.M. (1973). Adaptation to maximal effort: Genetics and age. *Journal of Applied Physiology*, **35**, 288-293.

Klissouras, V. (1972). Genetic limit of functional ability. *Internationale Zeitschrift Angewandte Physiologie*, **30**, 85-94.

Komi, P.V., Ito, A., Sjödin, B., Wallenstein, R., & Karlsson, J. (1981). Muscle metabolism, lactate breaking point, and biochemical features of endurance training. *International Journal of Sports Medicine*, **2**, 148-153.

Komi, P.V., & Karlsson, J. (1979). Physical performance, skeletal muscle enzyme activities, and fiber types of monozygous and dizygous twins of both sexes. *Acta Physiologica Scandinavica* (Suppl. 462), 1-28.

Komi, P.V., Viitasalo, T.J., Havu, M., Thorstensson, A., Sjödin, B., & Karlsson, J. (1977). Skeletal muscle fibers and muscle enzyme activities in monozygous and dizygous twins of both sexes. *Acta Physiologica Scandinavica*, **100**, 383-392.

Lesage, R., Simoneau, J.A., Jobin, J., Leblanc, J., & Bouchard, C. (in press). Familial resemblance in maximal heart rate, blood lactate and aerobic power. *Human Heredity*.

Lortie, G., Simoneau, J.A., Boulay, M.R., & Bouchard, C. (1986). Muscle fiber type composition and enzyme activities in brothers and monozygotic twins. In R.M. Malina & C. Bouchard (Eds.), *Sport and human genetics* (pp. 147-154). Champaign, IL: Human Kinetics.

Lortie, G., Simoneau, J.A., Hamel, P., Boulay, M.R., Landry, F., & Bouchard, C. (1984). Responses of maximal aerobic power and capacity to aerobic training. *International Journal of Sports Medicine*, **5**, 232-236.

Lortie, G., Bouchard, C., Simoneau, J.A., & Leblanc, C. (1983). La consommation maximale d'oxygène: Relations avec l'âge, le sexe, la graisse corporelle et le style de vie. In F. Landry (Ed.), *Health risk estimation, risk reduction and health promo-*

tion. Proceedings of the 18th annual meeting of the Society of Prospective Medicine (pp. 79-86). Ottawa: Canadian Public Health Association.

Lortie, G., Bouchard, C., Leblanc, C., Tremblay, A., Simoneau, J.A., Thériault, G., & Savoie, J.P. (1982). Familial similarity in aerobic power. *Human Biology*, **54**, 801-812.

Mahler, R.F. (1976). Disorders of glycogen metabolism. *Clinics in Endocrinology and Metabolism*, **5**, 579-598.

McKusick, V.A. (1982). Mendelian inheritance in man. *Catalogs of autosomal dominant, autosomal recessive, and X-Linked phenotypes*. Baltimore: Johns Hopkins University Press.

Montoye, H.J., & Gayle, R. (1978). Familial relationships in maximal oxygen uptake. *Human Biology*, **50**, 241-249.

Motulsky, A.G. (1977). Genetics of coronary heart diseases. *Pediatrician*, **6**, 366-370.

Pernow, B., & Saltin, B. (1971). Availability of substrates and capacity for prolonged heavy exercise in man. *Journal of Applied Physiology*, **31**, 416-422.

Personen, E., Norio, R., & Sarna, S. (1975). Thickenings in the coronary arteries in infancy as an indication of genetic factors in coronary heart disease. *Circulation*, **51**, 218-225.

Pollock, M.L. (1973). The quantification of endurance training programs. *Exercise and Sport Sciences Reviews*, **1**, 155-188.

Prud'homme, D., Bouchard, C., Leblanc, C., Landry, F., Lortie, G., & Boulay, M.R. (1984a). Reliability of assessments of ventilatory thresholds. *Journal of Sport Sciences*, **2**, 13-24.

Prud'homme, D., Bouchard, C., Leblanc, C., Landry, F., & Fontaine, E. (1984b). Sensitivity of maximal aerobic power to training is genotype-dependent. *Medicine and Science in Sports and Exercise*, **16**, 489-493.

Rost, W., & Hollmann, W. (1983). Athlete's heart: A review of its historical assessment and new aspects. *International Journal of Sports Medicine*, **4**, 147-165.

Rusko, H., Havu, M., & Karvinen, E. (1978). Aerobic performance capacity in athletes. *European Journal of Applied Physiology*, **38**, 151-159.

Rusko, H., Rahkila, P., & Karvinen, E. (1980). Anaerobic threshold, skeletal muscle enzyme and fiber composition in young female cross-country skiers. *Acta Physiologica Scandinavica*, **108**, 263-268.

Saltin, B. (1972). The effect of physical training on the oxygen transporting system in man. In G.R. Cumming, D. Snidal, & A.W. Taylor (Eds.), *Environmental effects on work performance* (pp. 151-162). Edmonton: Canadian Association of Sport Sciences.

Saltin, B., Blomqvist, G., Mitchell, J.H., Johnson, R.L., Wildenthal, K., & Chapman, C.B. (1968). Response to exercise after bed rest and after training. *Circulation* (Suppl. 7), 1-68.

Saltin, B., Henricksson, J., Nygaard, E., Andersen, P., & Jansson, E. (1977). Fiber types and metabolic potentials of skeletal muscles in sedentary man and endurance runners. *Annals of the New York Academy of Sciences*, **301**, 3-29.

Schantz, P., Billeter, R., Henriksson, J., & Jansson, E. (1982). Training-induced increase in myofibril ATPase intermediate fibers in human skeletal muscle. *Muscle and Nerve*, **5**, 628-636.

Tanner, J.M. (1964). *The physique of the Olympic athlete*. London: George Allen and Unwin.

Tesch, P., Daniels, W.L., & Sharp, S. (1982). Lactate accumulation in muscle and blood during submaximal exercise. *Acta Physiologica Scandinavica*, **114**, 441-446.

Weber, G., Kartodihardjo, W., & Klissouras, V. (1976). Growth and physical training with reference to heredity. *Journal of Applied Physiology*, **40**, 211-215.

Wilmore, J.H., Brown, C.H., & Davis, J.A. (1977). Body physique and composition of the female distance runner. *Annals of the New York Academy of Sciences*, **301**, 654-766.

4

Genetics of Health-Related Fitness

William J. Schull
UNIVERSITY OF TEXAS HEALTH SCIENCE CENTER
HOUSTON, TEXAS, USA

It grows increasingly clear that the common chronic diseases of middle and later life—atherosclerosis, cancer, diabetes, and hypertension—are the culmination of a long sequence of events having their origins in childhood (e.g., see Zinner, Margolius, Rosner, & Kass, 1976). Understandably, many of these cumulative events involve the so-called risk factors associated with the disease endpoints, that is, diet, lipid levels, life styles, and the like. Body composition, often measured simply as obesity, for example, looms large as one of the precursors of several of these disorders, yet many young athletes interested in weightlifting and some contact sports such as sumo wrestling or football are encouraged to achieve body weights and conformations that may be unhealthy. Surprisingly little is known about the normal changes in cardiovascular function, specifically blood pressure, in the teen and young adult years—those years from which most athletes are drawn. And even less is known about the atypical changes that may occur with unusual stress, weight gain, and the like imposed by rigorous training followed by relaxation of the training regime in subsequent years. These events are all under some degree of genetic control, the specific extent varying with the risk factor or factors involved.

Until quite recently, there has apparently been little interest in the role of genetic variation in athletic ability or human work performance in a more mundane sense (see Bouchard & Malina, 1983b; Bouchard & Lortie, 1984). Perhaps this is not unexpected, for work performance has tended to interest physiologists more than geneticists, and their perspectives are quite different. Physiologists generally seem preoccupied with averages, the prototypic performance if you will, not with variations about that average. But it is this variation that the geneticist seeks to understand.

At least two issues arise in furthering this assessment or understanding. There is first the matter of the exceptional athlete, the one whose performance shatters previous expectations, a Beamon, say, who moves forward the long-jump record by 2 feet whereas previously it had only inched ahead. Obviously, some of this is due to the honing of skills, though other long-jumpers had certainly honed their abilities no less zealously. Some may be due to exceptional environmental circumstances; thus in the instance cited, it is often asserted that the altitude in Mexico City with its lesser air density made this possible. However, other athletes have jumped at these altitudes and not shattered records. We are then left with an inexplicable residuum. Is this the result of a different genotype?

Second, a more pervasive issue is the matter of physical well-being and its relationship to disease, that is, health-related fitness. Whether genetic modulation in this instance is direct and the association with risk factors is fortuitous, or indirect and mediated through the risk factors themselves is usually not clear. However, the opportunities for the interaction of genetic and environmental factors seem virtually limitless, and the challenges to understanding greater.

Genes and Their Involvement in Health-Related Fitness

Cardiovascular disease, diabetes, hypertension, and often obesity are concomitants of socioeconomic change, or so it seems (Beaglehole, Eyles, Salmond, & Prior, 1978; Hanna & Baker, 1979; Marmot et al., 1975; McGarvey & Baker, 1979; Reed et al., 1982). The rapidity with which these diseases gain importance with such changes has generally been interpreted as evidence that environmental factors are of primary etiologic significance and that genetic ones are not. It is argued that "gene pools" cannot change swiftly enough to produce the differences that are seen. Indeed, this is generally true, but this is not the genetic question of moment, for this formulation of the issue assumes that a given genotype results in the same phenotype in every environment. However, this is commonly not so. Some environments optimize the phenotype associated with a particular genotype while others do not.

For example, a failure in the phloretin-sensitive lithium transport pathway generally imposes no handicap on most individuals but takes on special significance for the manic depressive who happens to inherit this defect and whose lithium therapy proves ineffective as a consequence (Ostrow, Pandey, Davis, Hurt, & Tosteson, 1978; Pandey, Ostrow, & Haas, 1977). Similarly, lactose intolerance is no burden to an individual in the absence of lactose, but in its presence it can be a debilitating if not life-threatening disorder to the genetically predisposed (Johnson, Kretchmer, & Simoons, 1974). The "thrifty genotype" argument so frequently invoked in discussions of noninsulin dependent diabetes mellitus and its rising prevalence in some populations is still another possible illustration (Neel, 1962). So too may be the situation concerning carriers of some genes associated with the inherited hemoglobinopathies when they reside in nonmalarious regions of the world. Are such effects to be considered environmental or genetic? Given the complexities of gene-environment interactions, any such simple assignment is surely specious and detracts from the real issues—the identification of high-risk individuals and through them the sequence of causal events.

Why should coronary heart disease, diabetes, cancer, and obesity be the diseases that increase? Why not just one or another? Or some other set? Is there a common biological intersection? And if so, what is it? At present none of these questions can be answered unequivocally, but new findings of a biochemical nature suggest common causal elements. Following is a loosely spun account of some of these findings, designed more to prompt reflection than to persuade.

Obesity

Obesity is possibly the most common of all deviations from the norm to be associated with disease. It has been said that no less than 27 health problems of consequence are directly related to being obese. Whether this is true may be moot, for the data are often conflicting, but there is certainly persuasive evidence that obesity is significantly associated with maturity onset diabetes (West, 1978), premature myocardial infarctions and hypertension (Pickering, 1968), cholecystitis and cholelithiasis (Abbruzzese & Snodgrass, 1970), cancer of the corpus of the uterus in women (Wynder, Escher, & Mantel, 1966), and gout (Wyngaarden, 1970). However, the chain(s) of events leading to these associations has been obscure. Some of the difficulties concern the very definition of obesity, which varies; most medical dictionaries are no more helpful than to describe it as fatness or corpulence. Often it is merely defined as some percentage excess over "the ideal body weight," another fuzzy parameter.

It has been common practice to treat obese individuals as if their obesity had the same causal basis, although experimental evidence has shown that there are at least two separable components in the deposition of excess body fat: hyperphagia and decreased metabolic efficiency. Each of these may in turn have heterogeneous origins, but this is conjectural since there has been no biochemical means to recognize such heterogeneity. However, this curtain of uncertainty may have been lifted at last. The relevant advances primarily involve the ionic pathways associated with the red blood cell membrane. These pathways are commonly termed active or passive; the distinction rests upon whether the movement of the ion follows the concentration gradient (a "passive" or diffusional pathway) or is counter to the concentration gradient (an "active" or nondiffusional pathway). Four pathways are presently recognized; there may be more. They include (a) the so-called Na-K pump, an active pathway that is inhibited by ouabain; (b) the sodium cotransport system, a passive pathway inhibitable by furosemide, a complex anthranilic acid often used as a diuretic; (c) the sodium-lithium countertransport system, another passive pathway inhibited by phloretin, a glucoside derived from the bark of many of the Rosaceae, and (d) the "leak," a minor pathway that seems to reflect a random loss of the ion in question.

York, Bray, and Yukimura (1978) have shown that an enzymatic defect exists in the obese mouse, a common animal model for obesity in man. There is a loss of thyroid-induced sodium- and potassium-dependent adenosine triphosphate; the homozygous obese mouse has reduced levels of sodium-potassium

ATPase. This observation prompted De Luise, Blackburn, and Flier (1980) and De Luise, Rappaport, and Flier (1982) to look for evidence of reduced energy use in the cells of obese persons since ATP is the principal energy currency of the body. They find the number of sodium-potassium pump units in the erythrocytes from obese subjects to be reduced about 22% as compared with nonobese controls, and also report the number of pump units to be significantly and negatively correlated with the percentage of ideal body weight.

Normally the sodium pump is responsible for 20 to 50% of total cellular thermogenesis (some estimates are as high as 70%). The obese individuals in this study weighed from 147 to 277% above their ideal body weight. They clearly satisfy most operational definitions of obesity but the selection criteria used may inadvertently compromise the inference which can be drawn, for the causal mechanisms that operate on the tails of a continuous distribution are often more limited in number than those seen more centrally in the distribution. It has been argued, for example, that some of the effect they observe is attributable to ethnic differences in obesity and that ethnic origin of patient and comparison person was not controlled in De Luise's studies (Beutler, Kuhl, & Sacks, 1983). Nevertheless, their observations suggest that within these obese individuals there exists some subset where the pump may be defective, cellular utilization of ATP diminished, and the calories that would normally find their way to glycolysis are stored.

Given the quantitative nature of De Luise et al.'s findings, one might assume that the $Na^+ - K^+$ pump is genetically controlled but suspect a multifactorial basis for the inheritance of differences. Most enzymatic variation appears continuously distributed, however, if measured in terms of activity levels even when the underlying structural differences are known to be simply and discretely inherited. The seemingly continuous nature of the variation in the pump may merely reflect the metric used to assess differences, and not the nature of the genetic variability.

While it is too early to know whether these findings will be supported with time, some observations suggest they may be. For example, there is evidence that some obese persons have an impaired thermogenetic response to a stimulus such as a rise in circulating catecholamines (Jung, Shetty, James, Barrand, & Callingham, 1979). Or, weight gain is a commonly observed phenomenon among manic depressives on lithium therapy, thought to be due to the known effects of lithium on water balance; polydipsia and polyuria are frequently seen in manics on lithium therapy. While this undoubtedly seems to be one mechanism that contributes to weight gain, there is also a gain in weight not attributable to retained water (Dempsey, Dunner, Fieve, Farkas, & Wong, 1976). Lithium impinges on the sodium-potassium pump and seems able to do so at lithium concentrations in the range of one milligram per liter of water (Clench, Ferrell, Schull, & Barton, 1981). Thus, weight gain with lithium therapy, and possibly at still lesser levels of lithium absorption, suggests a change in the sodium-potassium pump, either inherited or directly due to the competitive effects of this metal. These observations aside, a defect in the pump makes the widespread effects of obesity more easily understood.

There is of course other evidence of the role of genetic factors in human fatness. Some of this is biochemical, such as the apparent functioning of the enzyme, lipoprotein lipase, as the gatekeeper for the entry of lipids into the cell; some is not. Mueller (1983; see also Savard, Bouchard, Leblanc, & Tremblay, 1983) has recently reviewed this latter evidence and concludes that there is low to moderate heritability of adult static fatness. He asserts that approximately one-third of the variation between individuals in fatness seems due to genetic causes, and further notes that changes in fatness in the course of life as well as the anatomical positioning of fat are important modifiers of the health effects of obesity. Notable in this regard is the apparent centripetal distribution of fat seen in diabetes (e.g., see Joos, Mueller, Hanis, & Schull, 1984).

Blood Pressure Variability

Essential hypertension in adults is the single most common risk factor associated with cardiovascular disease; however, to understand the role of blood pressure variability in the underlying degenerative process(es) one must not only know about proximal events that may contribute to disease but about early ones as well. This includes the evolution of blood pressure in the young and the factors impinging on this evolution. Obviously, one of these is genetic variation. Pressure in the circulatory system is created by a combination of three basic factors (Guyton, 1981): the resistance of the peripheral vascular bed to the passage of blood through it, the total volume of blood forced through that bed, and the force created by the contraction of the heart.

However, the circulatory system has several characteristics that complicate the simple biophysical determinism of blood pressure; for example, body size and structure are variable, the heart is adaptable, and blood volume is adjustable. To the extent that such purely biophysical concepts encompass the phenomenon, one should be able to explain the observed variation in blood pressure by considering variation in more primary traits such as body size and composition, heart size and activity, and blood volume. While the evidence does suggest that blood pressure levels are associated with weight, it has not been possible to predict blood pressure values simply from data on weight, blood volume, and cardiac function. Other variables must be considered.

Psychosocial factors have often been invoked. Yet blood pressure does not rise or fall merely because of some poorly understood process called socioeconomic or psychosocial stress but because of hemodynamic changes, relaxation of the musculature of the circulatory vessels, and the like. Some biochemical sequence of events must intervene between the "change" or the "stress" (the trigger possibly) and the biophysical events ultimately manifested as an alteration in blood pressure. The challenge is to identify the latter; that is, delineate the biochemical events that result in the cellular or tissue changes reflected in altered blood pressure.

Significant advances have occurred in the understanding of the biochemistry and physiology of vasoregulation in humans. A variety of biological systems, including the renin-angiotensin-aldosterone system, the sympathetic nervous system, prostaglandins, and vasoactive peptides such as kinin appear to be involved in this phenomenon. A complete elucidation of all these systems and their interrelationships has proven elusive, however (see Weinshilboum, 1979, for a review of the biochemistry of hypertension).

The sympathetic nervous system and the human adrenergic neuron are illustrative of the interactions involved. Norepinephrine (noradrenaline) is the neurotransmitter of the adrenergic nerve terminal. Its action is normally terminated by a membrane uptake process of some complexity, but enzymatic metabolism is also involved. Two enzymes enter here—catechol-O-methyltransferase (COMT) and monoamine oxidase (MAO)—which operate on catecholamines of which norepinephrine is one. COMT catalyzes the methylation of endogenous catecholamines, including norepinephrine; whereas monoamine oxidase (inhibitors of this enzyme are used in the treatment of hypertension) catalyzes the conversion of norepinephrine to 3,4-dihydroxyphenylglycol aldehyde. The nature of the genetic control of these enzymes, that is, whether it is monogenic or polygenic, is not yet clear; however, twin and family studies have shown that variation in their activity seen in platelets (MAO) and erythrocytes (COMT) is inherited. As of yet, we know of no study published that attempts to correlate the activities of these enzymes to blood pressure or to obesity.

A relationship has long been suspected between sodium intake and hypertension (e.g., see Dahl, 1961; Morgan, Myers, & Carney, 1979). Anecdotal associations in humans were strengthened some years ago by the work of Lewis Dahl and his colleagues, who were able to develop two strains of rats, one of which always remained normotensive even on a high salt diet, whereas the other always became hypertensive. The salt-sensitive animals excreted a sodium load more rapidly than did the salt-resistant animals. However, studies of the correlation in human populations of blood pressure and individual salt intake have generally been inconclusive. Nonetheless, it is known that hypertensive individuals have an accelerated excretion of sodium when given a sodium load.

Recently, several abnormalities of sodium transport have been reported in the red blood cells of patients with essential hypertension and in a high proportion of their young normotensive offspring (e.g., Canessa et al., 1980; Garay, Dagher, & Meyer, 1980; Woods et al., 1982). These defects seem to involve the $Na^+ - K^+$ ATPase pump and the $Na^+ - K^+$ cotransport pathway. But the data are often conflicting. Garay and his colleagues, for example, have reported that some 50.6% of 97 normotensive offspring of parents, one or both of whom were hypertensive, exhibited the sodium-potassium cotransport defect, whereas 73.7% of 19 normotensive offspring of parents, both of whom were hypertensive, exhibited it. However, other investigators have been unable to confirm this (Etkin et al., 1982; Glynn & Rink, 1982; Hamlyn et al., 1982). Whether these differences reflect technical difficulties or something more fundamental has yet to be determined.

Another potentially informative ion channel involves the sodium-lithium countertransport system, which appears to have a bimodel distribution in population data (Turner, Johnson, Boerwinkle, Richelson, & Sing, 1985). Some 72% of the population appears to belong to the lower mode with a mean of 0.24 mmol x (liter RBC x h)$^{-1}$, and 28% to the upper mode with a mean of 0.42 mmol x (liter RBC x h)$^{-1}$. The upper mode of this distribution agrees closely with the mean seen in a sample of individuals with essential hypertension, and thus these individuals may represent a prehypertensive group, some or all of them destined to develop hypertension later in life. This seems to be a replicable distribution, for different investigators have obtained very similar means and variances for the two modal subsets as well as comparable fractions of the population distribution of Na-Li CNT activity in different populations.

The system is manifest early in life and appears consistent with single-gene inheritance (Canessa et al., 1980). It is important to note, first, that these differences do not involve the cellular content of Na, K, or Li but rather the rapidity of the movement of one or another of these cations across the cell membrane. This movement, insofar as it concerns the $Na^+ - Li^+$ countertransport, is twice as fast in hypertensives as in normotensives (Brugnara et al., 1983). Second, the flux difference between normo- and hypertensives appears restricted to those hypertensives with so-called essential hypertension; the flux rate does not differ between normotensives and those individuals whose hypertension is secondary to other diseases (Smith et al., 1984). Finally, after adjusting for weight, age, and sex, about 3% of all blood pressure variation in the normotensive population at large may be explained by variation in this system—but this might include a larger fraction of variation leading to essential hypertension (Boerwinkle, Turner, & Sing, 1984).

The Na-Li CNT system illustrates the value of searching for single gene systems that explain a fraction of hypertension and can be found in laboratory assays in the absence of blood pressure abnormalities. As previously indicated, other systems have been suggested for this kind of underlying physiological risk factors for hypertension, including the sodium-sodium ion transport system, and the renin system differences in blacks (e.g., Gillum, 1979). One possible way to identify the presence of still further such variants may be through the response to a direct challenge. This is in fact regularly occurring with drug therapies. Unfortunately, when a patient fails to respond or responds atypically to a drug, for example propanolol, the basis for this behavior is rarely pursued genetically. It is understandably easier to substitute an alternative medication.

In spite of the search for major genes, the genetic factors that determine blood pressure variation seem from available evidence to be largely of a polygenic nature (e.g., Hines, 1937; Krieger, Morton, Rao, & Azevedo, 1980). That is to say, single segregating Mendelian loci have not been shown to explain much of the observed family data, which is largely consistent with a sizeable component of variation in blood pressure being due to the cumulative action of many loci, each with small effect. This evidence stems from twin as well as family studies, to which we now turn briefly (see Siervogel, 1983, for a more measured review of the evidence than the brief one to follow, and Bouchard & Malina, 1983a, 1983b, for methodological issues that confront the genetic study of quantitatively varying traits such as blood pressure).

Most studies of the variability between monozygous and dizygous twins in blood pressure have focused on adults. Typical values for the intraclass correlation in systolic blood pressure in these circumstances are 0.55 for identical twins and 0.25 for fraternal twins (Feinlieb, Garrison, & Havlik, 1980; Hines, McIlhaney, & Gage, 1957; McIlhaney, Schaffer, & Hines, 1975). Commonly, fraternal twins are more similar than parent and offspring or siblings, although under a simple, additive genetic model these correlations should be the same. There is, or was of course, a shared environmental component in the one instance not present in the other. Correlations among juvenile twins are as high or even higher. Havlik et al. (1979), for example, found that monozygous twins identified in the Collaborative Perinatal Project of the National Institute of Neurological and Communicative Disorders and Stroke had intraclass correlations of over 0.5 for both systolic and diastolic pressures, as contrasted with values of 0.40 and 0.27 in these specific instances for fraternal twins. Presumably some of this higher correlation reflects a shared (still) environment.

Correlations of this magnitude give rise to high estimates of heritability, that is, the proportion of interindividual variability in blood pressure ascribable to genetic variation—substantially higher than those generally found in family studies. The latter fact prompted investigators at the University of Miami Medical Center to examine infant twins (Levine et al., 1980, 1982). They followed a cohort (67 monozygotic and 99 dizygotic twins) from birth to 1 year of age. Their estimates of heritability, based on blood pressure from 6 to 12 months of age, adjusted for gender and body weight, are 0.27 for systolic pressure and 0.17 for diastolic—values lower than those cited for adults. They cautiously conclude, "as regards the development of blood pressure, these data suggest that familial patterning reflects, in part, variation in genetic susceptibility to environmental factors" (1982, p. 764). It is hard to take exception to this conclusion, but it is also difficult to see how it advances our understanding of mechanisms.

Twin studies have their limitations; they provide little insight, for example, into segregation frequencies that must ultimately be determined from family studies. We cite only two of several types of studies of families selected randomly that can provide information on the segregation frequency or ratio; they have been chosen more to illustrate a type than to project a conclusion. The first comes in a variety of guises but generally involves whole communities (Tecumseh, Michigan [Higgins, Keller, Metzner, Moore, & Ostrander, 1980]; Muscatine, Iowa [Clarke, Schrott, Leaverton, Connor, & Lauer, 1978]), whole islands (Tokelau [Beaglehole, Salmond, & Prior, 1975]), and the like.

It is sometimes argued that these studies are free from sampling errors for they involve the "universe" of possible measurements. This argument is flawed, however. It assumes a steady-state, for otherwise a set of observations, even if exhaustive, is only a sample in time. Furthermore, participation is often incomplete, and the decision to be examined may not be random with respect to blood pressure. Rarely are all members of a family available for study, and so migrational effects need to be considered too. Nonetheless, the Tecumseh study shows that blood pressure in childhood and parental blood pressure are correlated with blood pressure levels in young adult life. This study shows measures of fatness to be correlated too, and it is not immediately clear whether the correlation in blood pressure values is secondary to the aggregation of other risk factors that cluster in families.

Hennekens and his colleagues (1976) in a somewhat different study found that blood pressures among siblings aggregate over the years 2 to 14. However, when newborn scores were substituted no aggregation was demonstrable, and from this they conclude that a familial influence on blood pressure in infancy is apparent by 1 month of age. In this as in most family studies of quantitatively varying characteristics, it is difficult to dissociate genetic effects from those that are ascribable to shared or correlated environments or both (e.g., see Hanis, Sing, Clarke, & Schrott, 1983a; Hanis et al., 1983b; Rose, Miller, Grim, & Christian, 1979).

A second type of family study that could contribute to further quantitation of sources of variation in blood pressure concerns families with both biological and adoptive children. Of such studies, the largest as yet reported involves the work of Biron and his colleagues (Annest, Sing, Biron, & Mongeau, 1979a, 1979b; Biron, Mongeau, & Bertrand, 1974, 1975; Mongeau, Biron, & Bertrand, 1977). They find that the correlation between foster parent and adoptive child is consistently smaller than the correlation between biological parent and natural child. Similarly, the correlation between natural siblings is greater than that between siblings of whom either one or both were adopted.

They conclude that a shared environment explains larger fractions of the parent-natural child and the full sib correlations for diastolic than for systolic blood pressure. Accordingly, it is argued that strategies to modify diastolic blood pressure variability through intervention on household environmental factors seem more promising. They estimate too that approximately 34% of the age adjusted variation between individuals in systolic blood pressure can be attributed to genetic variation. The use of families with adoptive children obviously requires further scrutiny, for more recently the placement of children was often conditional on such potentially confounded characteristics as income, ethnicity, and the like.

The evidence does not suggest a strong relationship between the blood pressure value in early and late childhood (e.g., Clarke et al., 1978), and it has often been suggested that the relationship with body size is a better one (e.g., Feinlieb et al., 1980; Voors, Webber, Frerichs, & Berenson, 1977). Nonetheless, the relationship between weight and blood pressure is also not strong in early childhood (Beaglehole et al., 1975; Feinlieb et al., 1980), though it seemingly becomes stronger later (Holland & Beresford, 1975). Similarly, the level of blood pressure and its rate of change are not consistently related (Feinlieb et al., 1980; Higgins et al., 1980; Hofman & Valkenburg, 1983; Zinner, Martin, Sacks, Rosner, & Kass, 1975). This suggests that there may be little predictive value, genetic or otherwise, to blood pressures taken early in childhood.

The age changes in blood pressure associated with childhood by and large are associated with growth and maturation, whereas the changes during adulthood are based on alterations in tissues due to use and exposure but not to programmed developmental change. The former are purely generative changes and the latter degenerative, and there is no a priori reason why these should be related. Most chronic diseases with both adult and childhood forms have different, distinct etiologies in their different forms, and often different pathologies and clinical courses.

Lipids

Forty years or so have elapsed since it was first apparent that substantial variability exists among individuals in their plasma lipids, that this variation is at least partly under genetic control, and that the risk of atherosclerosis can be associated with this variability. Wilkinson and his colleagues in a 1948 investigation of a large Michigan family with so-called essential familial hypercholesterolemia recognized the significance of the increase in total blood cholesterol in the occurrence of premature cardiovascular disease. On the basis of the cholesterol values within this family, they defined three phenotypes: a group of normocholesterolemic individuals; a group of heterozygous persons with elevated cholesterol values but free of clinical symptoms, more specifically, xanthoma; and a group of individuals homozygous for the gene which in the heterozygous state merely leads to increased blood cholesterol. Goldstein and Brown (1983) have subsequently shown that this disorder stems from an inherited defect in a membrane receptor, and that some 2% of the population of the United States carry this gene, generally in single dose.

Cholesterol and triglycerides, although unquestionably useful surrogates in the study of abnormalities in lipid metabolism, have not proven as revealing as a direct analysis of the various lipoproteins themselves. The latter can be separated ultracentrifugally on the basis of their densities in the presence of a flotation gradient. Four readily differentiable families of plasma lipoproteins are now recognized: (a) chylomicrons, (b) very low density or prebeta-lipoproteins (VLDL), (c) low density or beta-lipoproteins (LDL), and (d) high density or alpha-lipoproteins (HDL). Distinctions more complex than this are being proposed. Be this as it may, the HDL fraction seems influenced not only by diet but by exercise as well.

The latter may be particularly relevant to marathon runners. Cholesterol is the constituent of HDL most commonly measured, and there is compelling evidence that decreased plasma HDL cholesterol concentrations are associated with premature coronary artery disease. Assessment of risk is complicated, however, for the risks associated with various lipid components are variable with high LDL, a risk factor for coronary heart disease, and HDL, a protective factor.

It is now known that endogenously synthesized fat is carried from the liver in VLDL and that exogenous dietary fat is transported from its site of intestinal absorption in chylomicrons. These two triglyceride-rich particles are composed of a hydrophobic lipid core surrounded by a surface of polar lipids and proteins called apolipoproteins, and are metabolized through the action of lipoprotein lipase. Of late, research interest has centered on the activities of this enzyme and the apolipoproteins and the differences found in their content in chylomicrons and VLDL. The action of lipoprotein lipase leads to a diminution of the triglyceride core of these particles and to the transference of both the surface lipids and the apolipoproteins to HDL. The major constituents of the latter fraction, that is, of HDL, are apo A-I and apo A-II; among the minor constituents are the apolipoprotein B, Lp(a), C-I, C-II, C-III, D, E, F, G, and H.

Functionally, it is thought that apo A-I and C-I both activate lecithin-cholesterol acyltransferase and apo A-II enhances hepatic lipase activity. Lipoprotein lipase itself is activated by apolipoprotein C-II, and apo C-III inhibits the uptake in the liver of chylomicron remnants. Apo B containing particles bind to a specific LDL receptor on the cell surface, and upon internalization inhibit the activity of 3-hydroxy, 3-methylglutaryl (HMG) CoA reductase, the rate-limiting enzyme in cholesterol synthesis. A variety of apolipoprotein A-I variants have already been described and the nature of the amino acid substitution identified (see Schaefer, 1984, for a brief review).

While the full range of functions of the apolipoproteins has yet to be revealed, it is already clear that their roles are central to the transport of both exogenously derived and endogenously synthesized fat. It can therefore be presumed that their deficiency could have profound consequences. Indeed, this appears to be so, for inherited deficiencies are now known for apo A-I, apo B, apo C-III, and apo E, and even more complex deficiencies exist such as those associated with Tangier and fish-eye disease, or HDL deficiency with planar xanthomas (Schaefer, 1984). All of these phenotypes are rare; most appear to be the result of homozygosis and are generally associated with severe premature atherosclerosis and coronary artery disease. Collectively, of course, these genes are widely disseminated. Many are partially expressed in the heterozygous state, and it is not clear as yet whether the health of the heterozygotes is compromised. If the latter should prove true, and if it takes the form of premature coronary artery disease, then the health implications of the existence of these genes are formidable.

The simply inherited, monogenic forms of hyperlipidemia account, however, for a relatively small proportion of the hyperlipidemic individuals in our population (see Segal, Rifkind, & Schull, 1982, for a recent review). Most of the latter seem to stem from an interaction with the environment of numerous genes with small effects; that is to say, like variation in blood pressure, they appear to be quantitatively inherited (e.g., Family Study Committee for the Lipid Research Clinic Program, 1984; Green et al., 1984; Namboodiri et al., 1984a, 1984b; Siervogel & Glueck, 1980). Other factors are at work too. Exercise and the use of alcohol or drugs influence lipid levels. Identifying high-risk individuals early in life can be more difficult under these circumstances, and the efficacy of intervention more uncertain.

These complexities notwithstanding, the association of elevated cholesterol with increased premature atherosclerosis and cardiovascular disease has led to the promulgation of a "lipid hypothesis" which, in its simplest form, asserts that a reduction in lipid levels should be accompanied by a commensurate decrease in coronary artery disease. The recently published findings of the Collaborative Lipid Research Clinics of the National Heart, Lung and Blood Institute (USA) seem to bear out this prediction (Lipid Research Clinic Program, 1984a, 1984b).

Conclusions

As we have seen, a growing causal role is being recognized for seemingly simply inherited variation in such phenomena as sodium-potassium cotransport, sodium-lithium countertransport, and apolipoprotein variability. All of these are directly or indirectly associated with the production and utilization of adenosine triphosphate (ATP), the energy currency of the body. Ultimately, events that effect the cellular dynamics involved in the formation of the adenosine phosphates (mono, di, and tri) must impinge upon muscular strength and thus athletic performance since these are the sources of energy that fuel physical achievement. The membrane transport phenomena to which we have alluded exhibit levels of genetic diversity that would not have been seriously entertained until very recently. Whether this is advantageous or disadvantageous to health has yet to be established. However, the existence of such diversity suggests a number of cautions. First, it would be well for the sports rule makers to consider the implications of the changes they advocate on the health of the athletes involved in a sport, and to be conservative in those areas in which our information is still inadequate. Second, some of the enthusiasm exhibited for the surveillance of the use of drugs might be better transferred to efforts to identify those athletes, active or potential, who are genetically ill-disposed to the physical demands of specific sports. Finally, we should become more aware of ethnic and racial differences that may be associated with health-related fitness.

References

Abbruzzese, A., & Snodgrass, P.J. (1970). Diseases of the gallbladder and bile ducts. In M.M. Wintrobe, G.W. Thorn, R.D. Adams, I.L. Bennett, Jr., E. Braunwald, K.J. Issellbacher, & R.G. Petersdorf (Eds.), *Harrison's principles of internal medicine* (6th ed.). New York: McGraw-Hill.

Annest, J.L., Sing, C.F., Biron, P., & Mongeau, J-G. (1979a). Familial aggregation of blood pressure and weight in adoptive families. I. Comparisons of blood pressure and weight statistics among families with adopted, natural, or both natural and adopted children. *American Journal of Epidemiology*, **110**, 479-491.

Annest, J.L., Sing, C.F., Biron, P., & Mongeau, J.-G. (1979b). Familial aggregation of blood pressure and weight adoptive families. II. Estimation of the relative contributions of genetic and common environmental factors to BP correlations between family members. *American Journal of Epidemiology*, **100**, 492-503.

Beaglehole, R., Salmond, C.E., & Prior, I.A.M. (1975). Blood pressure studies in Polynesian children. In O. Paul (Ed.), *Second International Symposium on the Epidemiology of Hypertension: Epidemiology and control of hypertension* (pp. 407-419). Chicago: Yearbook Medical.

Beaglehole, R., Eyles, E., Salmond, C., & Prior, I.A.M. (1978). Blood pressure in Tokelauan children in two contrasting environments. *American Journal of Epidemiology*, **108**, 283-288.

Beutler, E., Kuhl, W., & Sacks, P. (1983). Sodium-potassium-ATPase is influenced by ethnic origin and not by obesity. *New England Journal of Medicine*, **309**, 756-760.

Biron, P., Mongeau, J-G., & Bertrand, D. (1974). Familial aggregation of blood pressure in childhood is hereditary. *Pediatrics,* **54**, 659-660.

Biron, P., Mongeau, J-G., & Bertrand, D. (1975). Familial aggregation of blood pressure in adopted and natural children. In O. Paul (Ed.), *Second International Symposium on the Epidemiology of Hypertension: Epidemiology and control of hypertension* (pp. 397-405). Chicago: Yearbook Medical.

Boerwinkle, E., Turner, S.T., & Sing, C.F. (1984). The role of the genetics of sodium lithium countertransport in the determination of blood pressure variability in the population at large. *Progress in Clinical and Biological Research,* **165**, 479-507.

Bouchard, C., & Lortie, G. (1984). Heredity and endurance performance. *Sports Medicine*, **1**, 38-64.

Bouchard, C., & Malina, R.M. (1983a). Genetics for the sport scientist: Selected methodological considerations. *Exercise and Sport Sciences Reviews*, **11**, 275-305.

Bouchard, C., & Malina, R.M. (1983b). Genetics of physiological fitness and motor performance. *Exercise and Sport Sciences Reviews*, **11**, 306-339.

Brugnara, C., Corrocher, R., Foroni, L., Steinmayr, M., Bonfanti, F., & De Sandre, G. (1983). Lithium-sodium countertransport in erythrocytes of normal and hypertensive subjects. *Hypertension*, **5**, 529-534.

Canessa, M.L., Adragna, N.C., Solomon, H.S., Connolly, T.M., Tosteson, B.S., & Tosteson, D.C. (1980). Increased sodium-lithium countertransport in red cells of patients with essential hypertension. *New England Journal of Medicine*, **302**, 772-776.

Clarke, W.R., Schrott, H.G., Leaverton, P.E., Connor W.E., & Lauer, R.M. (1978). Tracking of blood lipids and blood pressures in school age children: The Muscatine study. *Circulation*, **58**, 626-634.

Clench, J., Ferrell, R.E., Schull, W.J., & Barton, S.A. (1981). Hematocrit and hemoglobin, ATP and DPG concentrations in Andean Man. In G.F. Brewer (Ed.), *Proceedings of the Fifth International Conference on Red Cell Metabolism* (pp. 747-762). New York: Alan R. Liss.

Dahl, L.K. (1961). Possible role of chronic excess salt consumption on the pathogenesis of essential hypertension. *American Journal of Cardiology*, **8**, 571-575.

De Luise, M., Blackburn, G.L., & Flier, J.S. (1980). Reduced activity of the red-cell sodium-potassium pump in human obesity. *New England Journal of Medicine*, **303**, 1017-1022.

De Luise, M., Rappaport, E., & Flier, J.S. (1982). Altered erythrocyte Na^+-K^+ pump in adolescent obesity. *Metabolism*, **31**, 1153-1158.

Dempsey, G.M., Dunner, D.L., Fieve, R.R., Farkas, T., & Wong, J. (1976). Treatment of excessive weight gain in patients taking lithium. *American Journal of Psychiatry*, **133**, 1082-1084.

Etkin, N., Mahoney, J.R., Forsthoefel, M.W., Eckman, J.R., McSwigan, J.D., Gillum, R.F., & Eaton, J.W. (1982). Racial differences in hypertension-associated red cell sodium permeability. *Nature*, **297**, 588-89.

Family Study Committee for the Lipid Research Clinic Program (1984). The Collaborative Lipid Research Clinics Program Family Study. I. Study design and description of data. *American Journal of Epidemiology*, **119**, 931-943.

Feinlieb, M., & Garrison, R.J. (1979). The contribution of family studies to the partitioning of population variation of blood pressure. In C.F. Sing & M. Skolnick (Eds.), *Genetic analysis of common diseases: Applications to predictive factors in coronary disease* (pp. 653-673). New York: Alan R. Liss.

Feinlieb, M., Garrison, R.J., & Havlik, R.J. (1980). Environmental and genetic factors affecting the distribution of blood pressure in children. In R.M. Lauer & R.B. Shekelle (Eds.), *Childhood prevention of atherosclerosis and hypertension* (pp. 271-279). New York: Raven Press.

Garay, R.P., Dagher, G., & Meyer, P. (1980). An inherited sodium ion-potassium ion cotransport defect in essential hypertension. *Clinical Sciences*, **59**, 191a-193s.

Gillum, R.F. (1979). Pathophysiology of hypertension in Blacks and Whites: A review of the basis of racial blood pressure differences. *Hypertension*, **1**, 468-75.

Glynn, I.M., & Rink, T.J. (1982). Hypertension and inhibition of the sodium pump: A strong link but in which chain? *Nature*, **300**, 576.

Goldstein, J., & Brown, M. (1983). Familial hypercholesterolemia. In J.B. Stanbury, J. Wyngaarden, D.S. Fredrickson, J. Goldstein, & M. Brown (Eds.), *The metabolic basis of inherited disease* (pp. 672-712). New York: McGraw-Hill.

Green, P.P., Namboodiri, K.K., Hannan, P., Martin, J., Owen, A.R.G., Chase, G.A., Kaplan, E.B., Williams, L., & Elston, R.C. (1984). The Collaborative Lipid

Research Clinics Program Family Study. III. Transformations and covariate adjustments of lipid and lipoprotein levels. *American Journal of Epidemiology*, **119**, 959-974.

Guyton, A.G. (1981). *Textbook of medical physiology*. Philadelphia: Saunders.

Hamlyn, J., Ringel, R., Schaeffer, J., Levinson, P.D., Hamilton, B.P., Kowarski, A.A., & Blaustein, M.P. (1982). A circulating inhibitor of $(Na^+ - K^+)$ ATPase associated with essential hypertension. *Nature, 300*, 650-651.

Hanis, C.L., Sing, C.F., Clarke, W.E., & Schrott, H.G. (1983a). Multivariate models for human genetic analysis: Aggregation, coaggregation and tracking of systolic blood pressure and weight. *American Journal of Human Genetics*, **35**, 1196-1210.

Hanis, C.L., Ferrell, R.E., Barton, S.A., Aguilar, L., Garza-Ibarra, A., Tulloch, B.R., Garcia, C.A., & Schull, W.J. (1983b). Diabetes among Mexican-Americans in Starr County, Texas. *American Journal of Epidemiology*, **118**, 659-72.

Hanna, J.M., & Baker, P.T. (1979). Biocultural correlates to the blood pressure of Samoan migrants in Hawaii. *Human Biology*, **51**, 481-498.

Havlik, R.J., Garrison, R.J., Katz, S.H., Ellison, R.C., Feinlieb, M., & Myrianthopoulos, N.C. (1979). Detection of genetic variance in blood pressure of seven-year-old twins. *American Journal of Epidemiology*, **109**, 512-516.

Hennekens, C.H., Jesse, M.J., Klein, B.E., Gourley, J.E., & Blumenthal, S. (1976). Aggregation of blood pressure in infants and their siblings. *American Journal of Epidemiology*, **103**, 457-463.

Higgins, M.W., Keller, J.B., Metzner, H.L., Moore, F.E., & Ostrander, L.D. (1980). Studies of blood pressure in Tecumseh, Michigan. II. Antecedents in childhood of high blood pressure in young adults. *Hypertension*, Suppl. I, **2**, 117-123.

Hines, E.A., Jr. (1937). The hereditary factors in essential hypertension. *Annals of Internal Medicine*, **11**, 593-601.

Hines, E.A, Jr., McIlhaney, M.L., & Gage, R.P. (1957). A study of twins with blood pressures and with hypertension. *Transactions of the Association of American Physicians*, **70**, 282-287.

Hofman, A., & Valkenburg, H. (1983). Determinants of change in blood pressure during childhood. *American Journal of Epidemiology*, **117**, 735-43.

Holland, W.W., & Beresford, S.A. (1975). Factors influencing blood pressure in children. In O. Paul (Ed.), *Second International Symposium on the Epidemiology of Hypertension: Epidemiology and control of hypertension* (pp. 375-386). Chicago: Yearbook Medical.

Johnson, J.D., Kretchmer, N., & Simoons, F.J. (1974). Lactose malabsorption: Its biology and history. *Advances in Pediatrics*, **21**, 197-237.

Joos, S.K., Mueller, W.H., Hanis, C.L., & Schull, W.J. (1984). Diabetes Alert Study: Weight history and upper body obesity in diabetic and nondiabetic Mexican American adults. *Annals of Human Biology*, **11**, 167-172.

Jung, R.T., Shetty, P.S., James, W.P.T., Barrand, M.A., & Callingham, B.A. (1979). Reduced thermogenesis in obesity. *Nature, 279*, 322-323.

Krieger, H., Morton, N.E., Rao, D.C., & Azevedo, E. (1980). Familial determinants of blood pressure in northeastern Brazil. *Human Genetics*, **53**, 415-418.

Levine, R.S., Hennekens, C.H., Duncan, R.C., Robertson, E.G., Gourley, J.E., Cassady, J.C., & Gelband, H. (1980). Blood pressure in infant twins: Birth to 6 months of age. *Hypertension*, **2** (Suppl. I), 29-33.

Levine, R.S., Hennekens, C.H., Perry, A., Cassady, J., Gelband, H., & Jesse, M.J. (1982). Genetic variance of blood pressure levels of infant twins. *American Journal of Epidemiology*, **116**, 759-764.

Lipid Research Clinic Program. (1984a). The Lipid Research Clinics Coronary Primary Prevention Trial Results. I. Reduction in incidence of coronary heart disease. *Journal of the American Medical Assocation*, **251**, 351-364.

Lipid Research Clinic Program. (1984b). The Lipid Research Clinics Coronary Primary Prevention Trial Results. II. The relationship of reduction in incidence of coronary heart disease to cholesterol lowering. *Journal of the American Medical Association*, **251**, 365-375.

Marmot, M.G., Syme, L., Kagan, A., Kato, H., Cohen, J.B., & Belsky, J. (1975). Epidemiologic studies of coronary heart disease and stroke in Japanese men living in Japan, Hawaii and California: Prevalence of coronary and hypertensive heart disease and associated risk factors. *American Journal of Epidemiology*, **102**, 514-525.

McGarvey, S.T., & Baker, P.T. (1979). The effects of modernization and migration on Samoan blood pressures. *Human Biology*, **51**, 461-480.

McIlhaney, M.L., Schaffer, J.N., & Hines, E.A., Jr. (1975). The heritability of blood pressure: An investigation of 200 pairs of twins using the cold pressor test. *Johns Hopkins Medical Journal*, **136**, 57-64.

Mongeau, J.-G., Biron, P., & Bertrand, D. (1977). Familial aggregation of blood pressure and body weight. In M.I. New & L.S. Levine (Eds.), *Juvenile hypertension* (pp. 39-44). New York: Raven Press.

Morgan, T., Myers, J., & Carney, S. (1979). The evidence that salt is an important aetiological agent, if not the cause of hypertension. *Clinical Sciences*, **57**, 459s-462s.

Mueller, W.H. (1983). The genetics of human fatness. *Yearbook of Physical Anthropology*, **26**, 215-230.

Namboodiri, K.K., Green, P.P., Walden, C., Kaplan, E.B., Dawson, D., Kelly, K., Maciolowski, M., Morrison, J.A., Elston, R.C., Austin, M., Rifkind, B.M., & Thomas, R. (1984a). The Collaborative Lipid Research Clinics Program Family Study. II. Response rates, representativeness of the sample, and stability of lipid and lipoprotein levels. *American Journal of Epidemiology*, **119**, 944-958.

Namboodiri, K.K., Green, P.P., Kaplan, E.B., Morrison, J.A., Chase, G.A., Elston, R.C., Owen, A.R.G., Rifkind, B.M., Glueck, C.J., & Tyroler, H.A. (1984b). The Collaborative Lipid Research Clinics Program Family Study. IV. Familial associations of plasma lipids and lipoproteins. *American Journal of Epidemiology*, **119**, 975-996.

Neel, J.V. (1962). Diabetes mellitus: A "thrifty" genotype rendered detrimental by "progress"? *American Journal of Human Genetics*, **14**, 353-362.

Ostrow, D.C., Pandey, G.N., Davis, J.M., Hurt, S.W., & Tosteson, D.C. (1978). A heritable disorder of lithium transport in erythrocytes of a subpopulation of manic-depressive patients. *American Journal of Psychiatry*, **135**, 1070-78.

Pandey, G.N., Ostrow, D.C., & Haas, M. (1977). Abnormal lithium and sodium transport in the red cells of a manic patient and some members of his family. *Proceedings of the National Academy of Sciences, USA*, **74**, 3607-3611.

Pickering, G. (1968). *High blood pressure*. New York: Grune and Stratton.

Reed, D., McGee, D., Cohen, J., Yano, K., Syme, L.S., & Feinlieb, M. (1982). Acculturation and coronary heart disease among Japanese men in Hawaii. *American Journal of Epidemiology*, **115**, 894-905.

Rose, R.J., Miller, J.Z., Grim, C.E., & Christian, J.C. (1979). Aggregation of blood pressure in families of identical twins. *American Journal of Epidemiology*, **109**, 503-511.

Savard, R., Bouchard, G., Leblanc, C., & Tremblay, A. (1983). Familial resemblance in fatness indicators. *Annals of Human Biology*, **10**, 111-118.

Schaefer, E.J. (1984). Clinical, biochemical, and genetic features in familial disorders of high density lipoprotein deficiency. *Arteriosclerosis*, **4**, 303-322.

Segal, P., Rifkind, B.M., & Schull, W.J. (1982). Genetic factors in lipoprotein variation. *Epidemiology Reviews*, **4**, 137-160.

Siervogel, R.M. (1983). Genetic and familial factors in essential hypertension and related traits. *Yearbook of Physical Anthropology*, **26**, 37-63.

Siervogel, R.M., & Glueck, C.J. (1980). Interrelationships between lipids, lipoproteins and blood pressure in children from kindreds with essential hypertension. *Preventive Medicine*, **9**, 760-772.

Smith, J.B., Ash, K.D., Hunt, S.C., Hentschel, W.M., Sprowell, W., Dadone, M.M., & Williams, R.R. (1984). Three red cell sodium transport systems in hypertensive and normotensive Utah adults. *Hypertension*, **6**, 159-166.

Turner, S.T., Johnson, M., Boerwinkle, E., Richelson, E., & Sing, C.F. (in press). Distribution of sodium-lithium countertransport and its relationship to blood pressure in a large sample of blood donors. *Hypertension*.

Voors, A.W., Webber, L.S., Frerichs, R.R., & Berenson, G.S. (1977). Body height and body mass as determinants of basal blood pressure in children—the Bogalusa heart study. *American Journal of Epidemiology*, **106**, 101-108.

Weinshilboum, R. (1979). Hypertension, a biochemical approach. In C.F. Sing & M. Skolnick (Eds.), *Genetic analysis of common diseases: Applications to predictive factors in coronary disease* (pp. 157-181). New York: Alan R. Liss.

West, K.M. (1978). *Epidemiology of diabetes and its vascular lesions*. New York: Elsevier.

Woods, J.W., Falk, R.J., Pittman, A.W., Klemmer, P.H., Watson, B.S., & Namboodiri, K. (1982). Increased red-cell sodium-lithium countertransport in normotensive sons of hypertensive patients. *New England Journal of Medicine*, **306**, 593-595.

Wynder, E.L., Escher, G., & Mantel, N. (1966). An epidemiological investigation of cancer of the endometrium. *Cancer*, **19**, 489-520.

Wyngaarden, J.B. (1970). Gout and other disorders of uric acid metabolism. In M.M. Wintrobe, G.W. Thorn, R.D. Adams, I.L. Bennett, Jr., E. Braunwald, K.J. Isselbacher, & R.G. Petersdorf (Eds.), *Harrison's principles of internal medicine* (6th ed.) (pp. 597-605). New York: McGraw-Hill.

York, D.A., Bray, G.A., & Yukimura, Y. (1978). An enzymatic defect in the obese (ob/ob) mouse: Loss of thyroid-induced sodium- and potassium-dependent adenosine triphosphate. *Proceedings of the National Academy of Sciences, USA*, **75**, 477-481.

Zinner, S.H., Martin, L.F., Sacks, F., Rosner, B., & Kass, E.H. (1975). A longitudinal study of blood pressure in childhood. *American Journal of Epidemiology*, **100**, 437-442.

Zinner, S.H., Margolius, S.H., Rosner, B.R., & Kass, E.H. (1976). Does hypertension begin in childhood? Studies of the familial aggregation of blood pressure in childhood. In M.I. New & L.S. Levine, (Eds.), *Juvenile hypertension* (pp. 45-54). New York: Raven Press.

5

Genetic Determinants of Sports Performance

D.F. Roberts
UNIVERSITY OF NEWCASTLE UPON TYNE
NEWCASTLE UPON TYNE, ENGLAND

Sports performance is an attribute deriving both from human prehistory and from recent and highly artificial concepts. In the primate dawn, agility in climbing a tree out of the way of a predator or a dominant elder foreshadowed gymnastic ability. Primitive man's need for long and persistent pursuit of wounded game for the next day's food and for rapid bursts of acceleration and speed in escaping from a charging assailant foreshadowed the skills required in marathon running and sprinting. Size and strength were at a premium in struggles between competing individuals. On the other hand, the artificiality of modern pole-vaulting, shot-putting, competitive walking, skating, and so on, are of recent derivation. Success in the survival skills of early times would have led to natural selection that favored phenotypes enhancing that success. Its genetic component, however small, is likely to have been selected for. Yet to establish the precise extent of that genetic component is extremely difficult.

There are several reasons for this. The first is the long human generation length, which makes it difficult to obtain comparable records for parents, children, and other relatives. Second is the lability in performance that comes in the long term with aging and in the short term with variations in health and nutrition; and there is the problem of distinguishing the effects of intrinsic genetic characters from those due to environment. There are other complex biological features as well. An additional element, however, is the question of motivation and other psychological attributes, attitudes to sport, and the general environment conducive to development of the necessary skills. But most important is biological complexity itself. It is the almost infinite number of variables involved in athletic prowess that make it difficult to measure the genetic contribution to it.

Factors that Contribute to Athletic Prowess

Aerobic Power and Capacity

Bouchard's valuable examination of aerobic power and capacity follows others in his remarkable series on aspects of performance, for example on endurance (Bouchard & Lortie, 1984) and on physiological fitness and motor performance (Bouchard & Malina, 1983). He examines laboratory data on maximal aerobic power (MAP), which can be measured with a high degree of reliability, and the maximal aerobic capacity (MAC), for which he has succeeded in developing a reproducible and specific assessment. As a result, he has been able to examine ventilatory thresholds in prolonged aerobic exercise, submaximal power output in a relatively steady state, and determinants of aerobic power and capacity such as heart dimensions and skeletal muscle characteristics. Collecting data on nuclear families with biological children or adopted children, identical or fraternal twins, close relatives or more distant ones, and unrelated individuals living together or apart, and applying these data to standard methods of quantitative genetic analysis, has produced highly valuable information on the extent of genetic influence on these variables.

It is not surprising that MAP, MAC, and most of their determinants improve with training, and that trainability is limited. Such environmental effect extends from overall response down to the details of muscle oxidative enzyme activity. What is of interest, however, is that this training effect remains subordinate to the genotype, though the preliminary nature of the estimates are acknowledged. A curious finding that merits more attention is the difference in the correlation coefficients between mother/child and father/child in measured MAP. Skeletal muscle adapted to endurance has considerably more mitochondrial cristae per gram than untrained muscle, and the mitochondria, of course, contain oxidative enzymes. Since mitochondrial DNA codes for a large number of RNAs associated with the maintenance and function of the mitochondron, this may well be evidence of cytoplasmic inheritance—if one may speculate from such limited data. Concerning twin data on maximal aerobic power, it is particularly interesting that Bouchard's data point to appreciable heritabilities in maximal heart rate but less so in variables per unit body weight. Submaximal power output appears to have low but significant heritabilities of the determinants of MAP and MAC. The curious difference in the pattern of correlations between the left ventricular internal size measurements and the wall and septum thickness and mass requires explanation, and so does the difference between heritabilities of types I and IIa muscle fibers.

From Bouchard's findings in general, heritabilities tend to be moderate to low for maximal aerobic power, maximal aerobic capacity, and some of their determinants. Not surprisingly, the interactions appear complex. Age, previous experience, pretraining level, and sex all affect trainability and hence contribute to the environmental and interaction variances. Bouchard's work leaves little doubt that there are high and low responders to aerobic training, some responding early and some late, but there is still no genetic marker to predict the individual's sensitivity to training. However, the estimate that 5 to 10% of the population may be high responders gives some idea of the effort required to identify them.

From the pattern of correlations observed in the echographic measurements of heart dimensions is a little evidence that seems to give support. The two heart characteristics suggesting appreciable genetic effect are the left ventricular internal diameter and the left ventricular volume per unit body surface area. Three other characteristics that do not suggest genetic effect are left ventricular posterior wall thickness, the interventricular septum thickness, and the left ventricular mass. One way to identify genetic effects on a variable is by inquiring whether it shows up in inbreeding. This can only be done in a population characterized by frequent marriages between relatives, a wide range of inbreeding coefficients, and accurate pedigrees for inbreeding assessment.

Such a population is found on Tristan da Cunha, for which there is detailed pedigree information and measures of inbreeding on all individuals. Moreover, excellent ECG traces are available. In 34 males and 38 females in whom the electrocardiogram was completely normal, measurements on all traces were made blind, without knowledge of any other detail of the individual concerned. Three intervals were measured on the lead II trace (Table 1). The QT interval and the ST interval showed no relationship with inbreeding. However, the PR interval in both sexes showed a trend to diminution with increase in the inbreeding coefficient, and the regression was significant over the total sample. This trend was in the opposite direction to the effect on the PR interval of disease or drugs which depress conduction through the atrioventricular bundle. It was not due to reduced pulse rate, for there was no suggestion of any association of inbreeding with pulse rate. It seemed most likely to be genetic and suggested that recessive genes may be involved, distributed over a large number of loci.

Following the onset of contraction of the left ventricle, indicated by the Q wave, pressure rises within it, and closure of the mitral valve prevents return of the blood to the atrium. Pressure builds up rapidly in the left ventricle until it exceeds that in the aorta when the aortic valve opens and the blood in the left ventricle is expelled into it. The PR interval gives the conduction time from auricle to ventricle. It is during the PR interval that the left ventricular volume is in its later, slower phase of increase to maximum, so that the time taken to achieve filling is to some extent a function of internal size. By contrast, however, the ST interval is a period mainly of reducing left ventricular volume, and the QT interval of which it makes up the greater part, similarly. Therefore, the results of Bouchard's analysis and of ours seem to be compatible, suggesting that the genetic effect, modest though it is, is more pronounced in the left ventricular internal size. The Q and S deflections correspond to events during ventricular depolarization, and T to the spread through the ventricle of the repolarization wave associated with relaxation of the muscle contraction; to some extent, these are associated with the thickness of tissue. Again, the results in the Tristan study are compatible with those of Bouchard, suggesting less genetic influence on this.

Motor Development

Malina makes the important distinction initially that motor development and performance can be seen from two points of view: that relating to the processes by which it is brought about, and its product, the outcome of the activity.

Table 1. ECG intervals and inbreeding coefficients

		Males + Females	
Interval	F	n	mean
PR	0	13	0.169
	0.001 – 0.050	24	0.151
	0.051 – 0.100	28	0.149
	0.1	7	0.130
QT	0	13	0.363
	0.011 – 0.050	24	0.365
	0.051 – 0.100	28	0.361
	0.1	7	0.354
ST	0	13	0.282
	0.001 – 0.050	24	0.288
	0.051 – 0.100	28	0.285
	0.1	7	0.264

Most of the data deal with the products of movement, which is true also for attempts to estimate the genetic contribution to variance in motor tasks. His demonstration that twins show a consistent lag in motor development compared to singletons reminds us once again of the problems inherent in the use of twin studies to measure genetic influence, and indeed he is careful to note the reservations attaching to some of them—for example, omission of age differences beween twin pairs and inadequate zygosity estimates. The variation in estimates of heritability of strength measured in different ways is unfortunate, but in view of the technical difficulties it is not surprising. However, it seems that the heritability of static strength is in the middle of the range (approximately 50%), and the same appears true of measures of explosive and dynamic strength.

Similar variability is shown in estimates for balance, speed of limb movement, and dexterity and coordination. Assortative mating is another element to be taken into account in interpreting the intrafamilial correlations, and this has not always been done. A curious generalization is that brothers tend to resemble each other more than sisters do in strength and motor tasks, and this Malina attributes to environmental covariation which differs in sex. It is perhaps surprising to find a substantial genetic influence on throwing tasks. Clearly, a virtual infinity of tasks could be measured, and yet the great variation in estimates that occur when different researchers apply the same test points again to the need for delicate standardization as well as the need to control for cultural variables.

Malina's reference to perceptual motor characteristics and reaction time was of interest, particularly the contrast of low heritability for reaction time to a light stimulus in twin studies (0.22 and 0.55) to the high estimate (0.86) from another investigation, and an even higher estimate from patella reflex time. The results of our own work on nerve conduction velocity measurements are relevant, for although they range outside normal toward clinical biology, they show several important points.

During our studies on hereditary motor and sensory neuropathy in families presenting with the Charcot-Marie-Tooth phenotype, motor nerve function was examined in patients and their normal relatives. The terminal latency and conduction velocity were measured for the median nerve, ulnar nerve, and common peroneal, and their correlations between pairs of relatives of different degrees were examined (Table 2). The correlations between pairs of affected first-degree relatives first examined in the total series proved surprisingly high, with the overall levels diminishing from pairs of first-degree to second-degree to third-degree relatives. However, when the correlations were calculated taking into account differences between families, so that they were intrafamilial, the correlations disappeared, suggesting that they largely reflected the heterogeneity there is from one family to another in which this phenotype occurs. More important, the correlations were greatly reduced in pairs of affected by normal, and they virtually disappeared in the small series of pairs of normal by normal. It seems, therefore, that these correlations are a feature of the disease process itself, and that the variation in normal individuals in nerve conduction velocity may be too slight for useful genetic analysis by present methods of measurement.

In this study we were fortunate in having a substantial number of pairs of relatives available for analysis, considerably larger than many of the studies referred to by Malina. But the results suggest that in interpreting familial correlations, data should be carefully scrutinized to ensure that they do not include any subjects whose results are sufficiently removed from the main body of data as to influence appreciably the resulting correlations. Perhaps this is a factor contributing to the variability of the familial correlations that emerges from Malina's review.

Health and Fitness

Schull's thought-provoking contribution distinguishes between health and athletic prowess. He points out that risk factors known to predispose to chronic diseases in middle and later life may be enhanced in the course of preparing for certain athletic events. He disposes of the argument that the rapid increase in cardiovascular disease, diabetes, and other conditions indicates a nongenetic etiology by reminding us of the interaction between phenotype manifestation and environmental modification; change in the environment may reveal more or fewer individuals with given genotypes. He reminds us of the relevance of basic single-locus gene action in multifactorial systems, giving examples of the possible involvement of adenosine triphosphate and reduced cellular energy use in obesity, and the sodium potassium transport and sodium lithium countertransport systems in hypertension. The majority of monogenic characters that are known vary in frequency from one population to another, and any markers of sensitivity to training may well do likewise. Bouchard's statement that no markers yet identify sensitivity to training, combined with Schull's comments, point to an area that deserves intensive research. The frequency with which ankylosing spondylitis and other joint disorders occur in youngish men who were formerly rugby players adds support to his call for considering the long-term implications of athletic training.

Table 2. Correlations between relatives in nerve conduction velocity

	Overall						Within families			
	1st-degree		2nd-degree		3rd-degree		1st-degree		2nd-degree	
	r	df	r	df	r	df	r	df	r	df
Affected x affected										
Median nerve TL	0.654***	52	0.563***	41	0.332***	57	-0.149	37	0.036	35
CV	0.640***	52	0.420**	41	0.351**	57	0.060	37	-0.041	35
Ulnar nerve TL	0.687***	20	0.723***	16	-0.067	29				
CV	0.524***	16	0.852***	16	0.327	29				
Affected x normal										
Median nerve TL	0.009	38	-0.494	23	-0.282	47	0.256	27	0	17
CV	-0.350	38	-0.168	23	0.007	47	-0.013	27	-0.295	17
Ulnar nerve TL	0.144	17	0.171	8						
CV	-0.557	17	0.611	8						
Common peroneal TL	-0.079	9								
CV	-0.395	9								
Normal x normal										
Median nerve TL	0.038	6			-0.269	6				
CV	0.013	6			0.819	6				

Note. TL = terminal latency; CV = conduction velocity.
* Significant at .05%.
** Significant at .02%.
*** Significant at .001%.

Size, Shape, and Physique

Wilson's presentation reminded us of a primary factor in sports performance, namely body size. His analysis of data on over 400 pairs of twins followed from birth to 9 years, using standardized measures, demonstrates once again the similarity between monozygous pairs as opposed to dizygous pairs. The patterns of the growth curves imply steady replacement by the individual's own genes of the effects of size at birth and all the factors influencing it. As with all twin studies, the question of zygosity is critical. The probability of monozygosity depends upon the actual monogenic characters present in the twins, and some monozygotic pairs show low probabilities because of the high population frequencies of the blood groups and enzymes they possess. Certainly such pairs are likely to have occurred in Wilson's large sample, and it would be worthwhile to pick out all twin pairs in whom the probability of monozygosity is less than .99 and examine them on highly informative systems such as Gm or HLA types. The growth curves of the twins and the ways in which the members of a pair approach or diverge from each other at different ages are fascinating. Translating these descriptive curves into velocity curves, a more dynamic analysis would emphasize these movements and perhaps allow the identification of environmental variables (e.g., episodes of disease) which may not quite coincide in both members of a twin pair, for if so these would modify the attribution of such trends to genes coming into action at different times.

Perhaps most of the data on athletes relate to body size and physique and its genetic control, though this information is far from adequate as yet. Viewing the anthropometric data given by Tanner (1964) on Olympic athletes, for instance, one is impressed by the weight and stockiness of the shotput, weightlifting, and wrestling participants, and by the height and slenderness of the short-distance hurdlers and high-jumpers. Similar contrasts are seen when their physique is expressed by somatotyping, and Carter (1970, 1974) has shown how the distribution of exponents of particular sports tend to cluster in the somatocharts. But to appreciate the significance of these clusters, one must view them against the background of normal human variation, which may be examined at two levels. First are the differences that occur between populations, and second are those that occur among individuals within a population. Because of our preoccupation with differences between individuals for purposes of recognition and social interaction, we tend to think that such variation is more extensive than it actually is.

In my file of several thousand samples, the smallest peoples on record are the pygmy groups in central Africa. Gusinde (1948) found a mean stature of 143.8 cm for 386 Efe and Basua adult males, and 137.2 cm for 263 females. Data on the related Aka show a mean stature of 144.4 cm for 115 males and 136.7 cm for 110 females. Other pygmy groups outside Africa have mean statures several centimeters higher. The standard deviations in all pygmy groups average some 7 cm. The tallest peoples for whom I have records are the Northern Nilotes of the southern Sudan. For a sample of Ageir Dinka males, the mean is 182.9 cm with a standard deviation of 6.1 cm; for a sample of 227 Ruweng Dinka males the mean is 181.3 cm with a standard deviation of 6.4. The mean stature of the Ageir Dinka women is 168.9 cm, with a standard deviation of 5.8 cm. Thus, the greatest population mean is only 127% of the smallest for males, and only 123% for females. The tallest normal individual Dinka of whom I have record was 210 cm, whereas the smallest pygmy male was just under 130 cm, so the height of the tallest normal individual recorded is only 162% of the smallest.

The mean stature for some 1,200 samples from Africa and Europe is 167.1 cm, and the standard deviation of the means is approximately 5.6 cm. The standard deviation of stature is known for 200 of these samples. The mean standard deviation is 6.1 cm. In other words, variation in total body size between populations is, if anything, less than that within populations. Tildesley (1950) examined a number of anthropometric dimensions in a similar way for indigenous populations all over the world. In almost every dimension, the variability within populations was greater than that between them.

For body composition, data are far fewer. Indeed, the only variable on which there are beginning to be sufficient series relates to skinfold thickness. The people with the least mean subscapular skinfold are the Shilluk males (6.2 mm, SD 2.43), while among those with the largest mean subscapular skinfold are upper-class Mexican American males from San Antonio, Texas (22.0 mm, SD 7.7). The healthy person who had the least subcutaneous fat was an Ageir Dinka, with a skinfold profile of submandibular 2.6, left subscapular 4.9, left suprailiac 3.4, and left biceps 2.7 mm. The one who had the most was an obese young English woman upon whom it was impossible to persuade the calipers to bite at any site. The greatest population mean is 449% that of the smallest if measured linearly, but if in view of the distribution of the measurement it is assessed on a logarithmic scale, then the greatest mean is 335% that of the smallest. The standard deviation of the means for some 35 series of subscapular skinfolds with an overall mean of 9.6 mm is approximately 1.6 mm. Again the intrapopulation variation is greater than that of the interpopulation, and obviously, from the skewed nature of the intrapopulation distribution, the individual range to include the extremes would seem to be very great indeed. But allowing for this skewness by scaling, it seems that intrapopulation variation in such a body composition indicator, relative to interpopulation variation, may be slightly greater than in linear dimensions.

Thus there is ample information to show that in terms of size and shape, quantified by measurement and indices, the extent of such variation is considerable but not excessive. These population means are surpassed by some groups of athletes, for example high jump mean stature of 188.7 cm white and 191.5 cm negro (Tanner, 1964), but the individuals occur only in the tails of the distributions and very rarely at the extreme. It is not the very extreme individual who excels. A second point is that outstanding skill in some sports tends to be found more frequently in some populations than others—the gracileness so often found in Africans predisposes them to success in hurdles and high jump; yet the overlap between populations is such that rarely will an athlete from any one be precluded from excelling in any given event. The inter- and intra-group variabilities indicate that though a higher proportion of certain populations have the optimal physique for a given sport, there is sufficient variation within each population to provide individuals capable of competing well in almost any event.

How do we account for such variation in size, shape, and body composition? These may be due partly to the plasticity of development responding to environmental variation; undoubtedly they are brought about through differences in growth rate and duration as well as varying control of metabolic processes; and equally undoubtedly, the fundamental control is genetic though it interacts with the environment in which development occurs. So far, though there is a fair amount of information on size, shape, and physiological measures, and though the last few years have seen increasing attention to growth studies in different populations (of which Wilson has given a valuable illustration), there is very little information on genetic differences in the parameters of growth in human populations. From other genetic work, however, one can envision the nature of that genetic control.

Nature of Genetic Control

Whether an individual's red cells contain a particular type of enzyme (e.g., PGM 2-1 or PGM 1-1) or a certain type of hemoglobin is determined by the particular alleles at the controlling locus the individual inherits at conception. No matter what happens to him or her during life, that element of his or her body composition remains determined. Yet it only gradually evolves during ontogenesis at the appropriate key phases in embryonic or fetal or postnatal life.

The genetic information of the DNA that is transmitted on the parental chromosomes to the new organism at the moment of conception, and which is so accurately reproduced in subsequent cell divisions, controls the structures of all the proteins making up and made by the new organism; it regulates their synthesis and their interaction with other substances so that the biochemical makeup of each individual before and after birth is essentially a reflection of the genetic constitution. It is this biochemical makeup that in later life determines the distribution and amount of each of the major compartments of our body composition. Hence not only are inherited differences in normal physical, physiological, or mental characteristics likely to be a consequence of differences in enzyme or protein synthesis, but so too is the sequence of developmental differentiation and individual variations in it.

Of course, cell differentiation and development does not occur in a vacuum; it is a complex of gene-controlled synthetic processes proceeding in the environment provided by the cell cytoplasm, where it is subject to other influences such as from information in extra-chromosomal DNA or in other organelles of the cell. Indeed, some of the organelles into which the macromolecules are organized, especially the mitochondria and membranes, act as templates for their further assembly. Further instructions come via the highly organized, cyclical chemical reaction system, in which the product of one reaction forms the substrate for the next. Extra-chromosomal instructions appear to be especially important in the earliest stages of development immediately after fertilization. Later come the epigenetic mechanisms, response to position or developmental stresses.

The single cell that is the fertilized egg gives rise to different kinds of cells, different tissues, and different organs. This development is brought about by a sequence of changing populations of cells organized in increasingly complex patterns. Cytogenetics shows that from the behavior of the chromosomes at each mitotic cell division, daughter cells contain virtually identical sets of genes. Yet as division succeeds division, the cells diverge until they have evolved into types as different as those of muscle, pigment, or nerve. Visibly quite different in appearance and function, the cells appear to behave as though they had received different sets of genes at earlier cell divisions. This is clearly not so. It is not the genes or the chromosomes that are distributed unequally. Rather it is that the cell responds to only a small fraction of its genes in the differing chemical environments that have developed. In other words, the initial cells have far greater potential than they express.

The hemoglobins provide a model of how genetic control of such development may be exercised. The globin of the normal adult hemoglobin molecule consists of two pairs of polypeptide chains, that is, two α chains and two β chains. A different gene specifies each chain, in which the sequence of constituents, respectively, 141 and 146 amino acid residues, is controlled by the DNA base pair sequence. To each chain is attached a heme group so that the molecular weight of the total molecule is approximately 66,000. There is also a minor hemoglobin fraction (HbA$_2$) consisting of two α chains and two δ chains, the latter again differing in structure from the α and the β. But these hemoglobins do not persist throughout the whole life span. There is instead a developmental sequence. During the first 10 weeks of embryonic life there exist unique hemoglobins which contain quite different polypeptide chains (ϵ and ζ). The principal ones are Gower$_1$, Gower$_2$, and Portland. During early embryonic development at 4 to 5 weeks gestation α and γ chains begin to be synthesized, and there is a decrease in ϵ and ζ chain production, giving rise to the principal fetal hemoglobin F (α_2, γ_2) which is gradually replaced by the adult form (α_2, β_2) during later fetal and early postnatal life. These chains are synthesized in different tissues. This means that since all cells contain identical genetic information, specific genes must be switched on and off at the appropriate stage of development. This appears to be a fundamental developmental control system, the sequential switching on and off of specific genes.

Evidence of Genetic Control

Single Gene

The most detailed genetic control of body composition exists at the finest level of analysis. Many variant enzymes and various types of protein that make up a cell are known, and at the cellular level the combination that occurs in a given individual is indeed virtually unique. Many such variants are Mendelian, that is, controlled by a single pair of genes, whereas in others the situation is more complex and it is their subunits that are so controlled. But at the intracellular level there is strong evidence for an intricate and precise control of body composition. However, when we look at the variables included in most body composition studies—relative amounts of adipose, bone, and muscle tissue, bone mineral content, and body water—these are at a much grosser level, body composition in the round as it were.

For such complex variables, strict and simple genetic control is neither seen nor expected. Yet one can still pick out single genes that affect body composition at this gross level, mainly those observed in Mendelian disorders. For example, the massive rotundity and fatness of the Laurence-Moon-Bardet-Biedl syndrome is part of the effects of the pair of alleles at the locus controlling this autosomal recessive disorder. Similarly, the skeletalization in the Cockayne syndrome, perhaps not so obvious at birth and in early infancy but becoming apparent in mid-childhood, is again a manifestation of autosomal genes producing this recessive condition. That there may be single genes relevant to athletic performance Schull has already noted.

Chromosomal

Chromosomal imbalance is a second factor affecting body composition. For example, the curious amount and distribution of body fat, the pseudofeminine appearance about the hips and thighs, the gynaecomastia frequently so severe as to require cosmetic surgery, is very characteristic of Klinefelter's syndrome, which has the 47 XXY karyotype instead of the normal 46 XY. Lean tallness characterizes the 47 XYY, and short adiposity the isoX variant of Turner's syndrome (45 XO). Down's syndrome, trisomy 21, shows the characteristic disturbance of body proportions. These syndromes are not likely to be important in competitive sport, however.

In terms of athletic prowess, one chromosome pair that is critical is the sex chromosomes. If an individual is conceived with two Xs, even at peak form she will never be able to compete with an individual at peak form conceived with one X and one Y, for the different endocrine functions that are programmed by the genes on these chromosomes have a profound effect on body form, physique, musculature, strength, and so forth. This is of course recognized empirically in the sex division of competitive events. This is what lies behind the chromosome testing of competitors in female Olympic events to check whether their outstanding performance is associated with the presence of a Y chromosome.

Multifactorial

From the composite nature of most body composition variables—the way in which they must result from a combination of metabolic processes interacting with variations in nutrition, activity, and other factors, as with so many other quantitative body features—normal variations in body composition seem best explained on a multifactorial hypothesis in which the genetic contribution is polygenic. Polygenes are transmitted in the same way and in accordance with the same laws as major genes, but their effects do not provide sufficient discontinuity for individual study. A polygene acts as one of a system, the members of which may act together to effect large phenotypic differences or against each other so that similar phenotypes develop from different genotypes. An individual polygene has only a slight effect; it is apparently interchangeable with others within the system. It does not have an unconditional advantage over its allele, since its advantage depends on other alleles present in the system, and it cannot therefore be heavily selected for or against (i.e., not until its associates in the same system come to be linked with it).

In polygenic inheritance, similar phenotypes develop from different genotypes, and therefore great genetic diversity and potentiality for change is concealed behind the phenotypic variation. The situation is well exemplified by reference to stature. The result in the individual depends more on the number of genes for tallness that he or she carries than on the particular genes present. Figure 1 shows the variation that would be produced in a quantitative character if three pairs of allelic genes were responsible for it, the gene for larger size being denoted by the capital letter in each case. The diagram assumes that the genes are all equal in their effects, that they simply add to each other, and that the alleles of each pair are equally frequent. With more genes, the irregularities of the steps of the histogram disappear and the curve becomes smooth. Thus, a normal distribution curve of a quantitative character in a population may be attained entirely by its genetic determination. If such a population were uniformly exposed to a different environment increasing the character, the whole curve would shift to the right. If only part of the population were so exposed, the shift of the mean would be less but the variability of the curve would increase (Figure 2).

The extent of the genetic contribution to such quantitatively varying characters can be measured by partitioning the variance. In essence, the total phenotypic variance is divided into that due to (a) additive polygenic effects, (b) environmental effects, and (c) other sources. The additive genetic contribution expressed as a proportion of the total variance is known as the heritability. A character that is totally genetically determined would have a heritability of 100%; that for which all the variation was environmental would have a heritability of zero. Heritability can be measured in several ways, for example by examining the resemblance between relatives of different degrees or from twin studies. For stature the evidence points to a heritability of 56% in West Africa (Roberts, Billewicz, & McGregor, 1978) and to slightly higher estimates in Europe (Huntley, 1966: 58-92%) and America (Byard, Siervogel, & Roche, 1983: 68%), reflecting the importance of the environmental con-

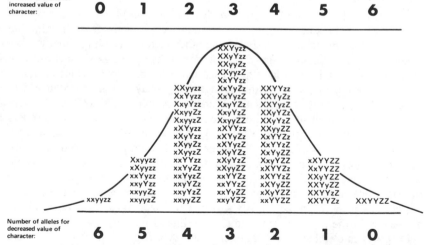

THREE LOCI, TWO ALLELES AT EACH, EQUAL AND ADDITIVE IN EFFECT: DISTRIBUTION OF GENOTYPES

Figure 1. Distribution of genotypes for three loci with two alleles at each loci. The example assumes that loci and alleles are equal and additive in effect.

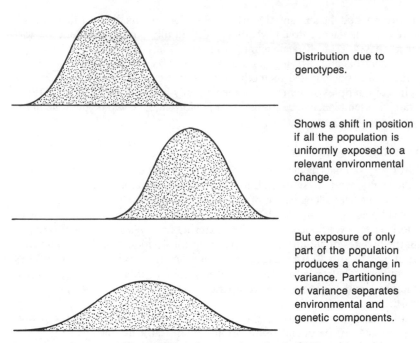

Distribution due to genotypes.

Shows a shift in position if all the population is uniformly exposed to a relevant environmental change.

But exposure of only part of the population produces a change in variance. Partitioning of variance separates environmental and genetic components.

Figure 2. Assumed distribution of genotypes (top). Shift produced by exposure of the entire population to a relevant environmental change (center). Distribution of characteristics when only part of the population has been exposed to the environmental change (bottom).

tribution in the harsh West African setting, perhaps due to the rigors of life and their impact on the population.

This model shows another important point. Even if the genetic contribution to variation within a population is said to be 100%, this implies nothing about the genetic basis of variation between populations. Theoretically it is possible for variation within a population to be totally genetic and for variation between populations to be purely environmental. It is not possible, therefore, to quantify the genetic contribution to any dimensional differences between populations, for this would require an analysis of hybrids. All that can be said is there is a high probability of an appreciable genetic contribution to the striking differences among peoples. The variation in height between the tall Nilotes and the small pygmies must derive essentially from genetic differences, for no pygmy would grow into a Nilote whatever the environment. A genetic basis to population differences in size, that is, differing frequencies of genes for tallness and shortness, seems reasonable in view of the considerable differences in gene frequencies that are well established for human populations for monogenic characters such as blood groups and isoenzymes.

It should be noted that most applications of this model in the performance studies discussed so far have been at the elementary level of twin and nuclear family analysis, which may not necessarily provide the most accurate estimates of heritability. This is not intended as a criticism; indeed, one has to start somewhere, and the amount of work already done in this difficult field is truly impressive. In fact, I doubt whether a full family study of athletic performance—based on an adequate number of propositions properly ascertained with proper controls and carrying out a number of tests and analyzing the results by the more sophisticated methods now available (e.g., path coefficients as applied to multiplex systems as in a recent study by Devor & Crawford, 1984, of family resemblance in neuromuscular performance)—will be possible in the immediate future.

Evidence From Growth Studies

Wilson also reminds us that athletes are adults or near-adults who have attained all the anatomical, physiological, and psychological elements making for their present skills over a long period of growth. Their adult bodies and achievements are visible summaries of all the events that have occurred during the growth period. These events include effects of external influences but they also represent the dictates of an intricate genetic program. Evidence of the genetic control of human growth from a steadily accumulating number of family studies is summarized elsewhere (Roberts, 1981), but in this context it is worth remembering that genetic control operates throughout the whole process of growth.

This is shown by evidence on growth landmarks such as age at peak velocity, at appearance of particular ossification centers, at eruption of teeth, at the occurrence of menarche, and at the attainment of various stages of the secondary sex characters. It is shown in ossification and dental sequences and timing, both in family studies and in racial comparisons. It is shown in the rates of growth at different ages, the development of different tissues at different ages, and in the shape and age-position of individual growth curves. Indeed, mathematical dissection of the last indicates that for given environmental circumstances the genetic control of the growth process extends down to many details of the velocity and acceleration curves. Moreover, there is evidence of the independence of genes. The genes controlling the rate of growth appear to be partly independent of those controlling final size, and genes concerned with height growth are not those concerned with adolescence. With very few exceptions, genetic control over most aspects of normal growth is multifactorial, but there is evidence of single locus effects in some sequence polymorphisms, for example in dental eruption and appearance of ossification centers.

In studies of the genetics of growth, similar difficulties occur and criticisms apply as to those of athletic performance. Yet the limited evidence there is is relevant. In view of the development patterns of body composition, individuals with an apparent propensity for physical performance should not wait until adulthood to train if optimal tissue proportions are to be achieved. Yet the earlier such training starts, the less is known about subsequent genetic growth programming, and insofar as the critical body proportions and size are finally determined only late in growth, some of those to whom this effort will be devoted will not develop the optimal morphologies. Neither is it known how the environmental modification of body tissues and physiological capacity through training may affect the implementation of genetic instructions that are to come later. Certainly in moderation it does not seem to be deleterious.

Several experiments with animals (e.g., Parizkova, 1975, 1983) indicated changes in the proportion of lean fat-free body mass by daily exercise and an increased ability to utilize lipid metabolites as fuel for muscle work. Similar studies in children taking regular moderate to intensive exercise showed increased lean body mass, aerobic capacity and heart size, and reduced fatness. In contrast, there appeared to be no significant effect on stature, skeletal maturation, and sexual maturation, although some data suggest both an accelerating and retarding effect of intensive training on sexual maturation of young athletes (Malina, 1983a, 1983b). In general, however, moderate encouragement of exercise seems advantageous.

General Points

While the physique of athletes has attracted much attention, this is but an external summary of the fundamental anatomical variables. The size, structure, insertions, and attachments of muscles have considerable effects on their mechanical efficiency, which itself depends on the efficiency of the cellular basis of muscular work, the contractile process, and the energy available for this. The change in rate of energy transformation is truly remarkable. If the resting metabolic rate of a man weighing 70 kg is about one kilocalorie per minute, on heavy exertion this may rise 20-fold to 20 kc per minute (1.3 kw), and there may be peaks in this over twice as high. It is therefore not necessarily the resting state on which selection acts, but the occasional peaks. To maintain such exertion there must be fuel and oxygen to provide the energy, and there must be efficient removal of the waste products of metabolism, carbon dioxide, and heat from the muscles involved and from the body as a whole. It may be necessary to channel oxygen from other organs to the muscles, for the equilibrium between accumulation and dissipation of metabolites is critical not only to the performance of the activity but also to the avoidance of pain and more serious tissue damage.

The cardiovascular system is of enormous importance. For a man weighing 70 kg, a large proportion of his 20 kg of skeletal muscle mass may be utilized in strenuous exercise, and the muscle blood flow may rise 12-fold or more. This is too much to achieve through blood flow redistribution, since other organs and especially the brain must continue to be supplied; it requires a 5- or 6-fold increase in cardiac output, for it is the maximum cardiac output as well as its distribution that largely determines the capacity for aerobic work. This increase is brought about by many factors: by the interaction or release of vasodilator substances from the working cells, a fall in the tissue tension of oxygen, and molecular splitting with the resultant release of energy. Stimulation of the adrenal medulla releases adrenalin and this contributes to the flow of a sympathetic alpha-noradrenergic vasoconstrictor, which assures the maintenance of blood pressure and redistributes the cardiac output to the dilated beds of the working muscles, though in stress situations this vasoconstrictive action may be effectively overriden by the metabolites from the molecular splitting. In submaximal work, increased cardiac output comes mainly with increased heart rate as a result of reduced parasympathetic and increased sympathetic drive. But when the effect of the autonomic nervous system on the heart is blocked so that normal rate changes are impossible, increased cardiac output is achieved by a bigger stroke volume. Hence, in enquiring into the genetic basis of athletic ability, one needs to examine the genetics not only of normal but also of reserve mechanisms.

The mechanisms in the many body systems and their interactions by which the human body maintains its normal activities are indeed intricate. The complex ways in which it stretches them to meet the demands of severe physical exercise are such that we may never be able to dissect out the genetic contribution to each.

Summary

The fundamentals of athletic performance are likely to have been selectively advantageous in prehistoric times, but it is difficult to identify the genetic contribution to them. Studies of two major physiological compartments of performance, motor development and aerobic power and capacity, show that they comprise components of heritability varying from low to moderate; none is sufficiently high to suggest that response to selection would be rapid or that outstanding athletic ability in parents would confer the same in any child.

In physique, as assessed by somatotyping, first-class performers tend to form clusters; this indicates there are physiques that are apparently optimal for a given event, and that these vary from one event to another. In body measurements, outstanding performers in some events tend to occur at the tails of the distributions of anthropometric traits but not at the extremes. Because of normal population differences in physique and body size, outstanding athletic skill for particular events is likely to be found more frequently in some populations than in others.

The question of single genes being relevant to athletic performance is raised; certainly there is chromosomal influence, but in most of the variables reviewed the genetic contribution is likely to be multifactorial. The fundamental developmental control system appears to be through sequential switching on and off of specific genes. The characteristics of adult athletes derive from events during the growing period; there is evidence that genetic control operates throughout the whole process of growth, and performance may be seen as the outcome of interaction between genetic predisposition and environmental modifications both at the adult and pre-adult stages through training.

Added to the difficulties of studying the genetic basis of sports performance that are common to other biological characters in humans, due to the long generation length and short-term variations in the character measured, are the following: First, selection is more likely to have acted upon the occasional peaks of exertion rather than on the resting state; second, such peaks may be attained by calling into play reserve physiological mechanisms, and it is these that must be considered in any study of the genetic basis of athletic ability.

The complexity of the biological basis of athletic performance makes this a singularly challenging area of investigation. To dissect out various contributory components in athletic prowess and examining one by one the genetic contribution to them seems the most promising line of attack.

References

Bouchard, C., & Lortie, G. (1984). Heredity and endurance training. *Sports Medicine*, **1**, 38-64.

Bouchard, C., & Malina, R.M. (1983). Genetics of physiological fitness and motor performance. *Exercise and Sports Science Reviews*, **11**, 306-339.

Byard, P.J., Siervogel, R.M., & Roche, A.F. (1983). Familial correlations for serial measurements of recumbent length and stature. *Annals of Human Biology*, **10**, 281-293.

Carter, J.E.L. (1970). The somatotypes of athletes—A review. *Human Biology*, **42**, 535-569.

Carter, J.E.L. (1974). Physical anthropology of the athletes. In A.L. de Garay, L. Levine, & J.E.L. Carter (Eds.), *Genetic and anthropological studies of Olympic athletes* (pp. 27-82). New York: Academic Press.

Devor, E.J., & Crawford, M.H. (1984). Family resemblances for neuromuscular performance in a Kansas Mennonite community. *American Journal of Physical Anthropology*, **64**, 289-296.

Gusinde, M. (1948). *Urwaldmenschen am Ituri*. Vienna: Springer-Verlag.

Huntley, R.M.C. (1966). Some problems in the study of quantitative variation in man. In G.G. Spickett & J.G.M. Shire (Eds.), *Endocrine genetics* (pp. 229-246). Cambridge: Cambridge University Press.

Malina, R.M. (1983a). Human growth, maturation, and regular physical activity. *Acta Medica Auxologica*, **15**, 5-27.

Malina, R.M. (1983b). Menarche in athletes: A synthesis and hypothesis. *Annals of Human Biology*, **10**, 1-24.

Pǎrizkova, J. (1975). Functional development and the impact of exercise. In S. R. Berenberg (Ed.), *Puberty* (pp. 198-219). Leiden, The Netherlands: Stonfert Kroese BV.

Pǎrizkova, J. (1983). *Growth, fitness and nutrition in pre-school children*. Prague, Czechoslovakia: Charles University Press.

Roberts, D.F. (1981). Genetics of growth. *British Medical Bulletin*, **37**, 239-246.

Roberts, D.F., Billewicz, W.Z., & McGregor, I.A. (1978). Heritability of stature in a West African population. *Annals of Human Genetics*, **42**, 15-24.

Tanner, J.M. (1964). *The physique of the Olympic athlete*. London: Allen & Unwin.

Tildesley, M.L. (1950). The relative usefulness of various characters on the living for racial comparison. *Man*, **50**, 14-18.

6

Heredity and Psychomotor Traits in Man

Napoleon Wolański
POLISH ACADEMY OF SCIENCES
WARSAW, POLAND

When speaking of selection for sport, two considerations are central. One can to some extent, be referred to as preselection. It should be based on knowledge of physique, physiological and psychomotor traits specific to attaining excellence in sport. However, it must be recognized that champions do not always exhibit these specific traits. The second consideration is performance, and the setting of performance records is complex. Records are sometimes attained by individuals with an average physique or average physical capacity, but usually only after long-term training. Such records are the sum of inborn characteristics, work, and "good fortune." Good fortune consists of happening upon an individual's high physical and psychic ability at the time a record is set. This may be related to biorhythms, but overall external conditions must also be considered. Nevertheless, it can be demonstrated that individuals with psychophysical traits typical of a given sport or event within a sport have a greater chance of attaining outstanding results.

This is associated with the first consideration, preselection for sport. Preselection, of course, indicates that selection has been made long before the individual is able to master a given sport. This may occur in early childhood, at the so-called golden age of motoricity (i.e., 4 to 6 years of age), in order to provide as much stimulation as possible for the desired traits. However, it is difficult to predict individual development. Therefore, a second stage of selection follows. It usually occurs spontaneously during training when less talented or less persistent individuals drop out. The second stage thus focuses on the selection of individuals achieving the best results and making the best progress.

The potential capacity of a population has two sources—genetic factors and cultural traditions—both being the heritage which transmits these properties from one generation to another. Both pathways are very conservative, and their effects are difficult to distinguish. This is one of the methodological difficulties in studies of heritability. There is at present a heated controversy between sociobiologists and their antagonists concerning the extent to which the gene pool determines not only reproduction of the population but also social development of the community. According to recent studies, predispositions to certain occupations, including sports, probably occur very early in life, at least in a general sense, although they are not always easily recognizable.

Although there is a relative abundance of materials on the heredity of psychomotor traits, there are methodological limitations. Two methodological approaches are chiefly used, one utilizing trait similarity between relatives and the other utilizing the variance occurring in a population or in certain cohorts with an expected degree of gene pool similarity.

Family Studies in Poland

This report presents an overview of our family studies in Poland. Details of the methods and sampling as well as an overview of the related literature have been previously reported (Wolański, 1973, 1984a; Wolański & Pyżuk, 1973; Wolański, Tomonari, & Siniarska, 1980). The available data consistently show that the size and shape of bones are more genetically controlled than soft tissue dimensions. A general interpretation of the coefficients of correlation between parents and offspring indicates that somatic traits are not necessarily more strongly determined genetically than psychomotor traits. Our data (Wolański & Kasprzak, 1979) indicate stronger genetic control of the blood levels of some enzymes (haptoglobin, MDH), and of grip strength and simple reaction time. For most psychomotor traits, the strength of genetic determination is medium (Figure 1): Speed, movement coordination, and endurance fitness appear to be more genetically controlled than the strength of large muscle groups and balance. Genetic control of the size of the head and body is similar. There is no clear-cut genetic determination of movement accuracy, but this may be due to technical difficulties in measurement of this trait.

The observation that genetic control of respiratory and circulatory traits is greater at work than at rest is important. Moreover, the genetic control appears to be stronger for the reflex-nervous-phase of work, that is, during the beginning and interruption of work, than for the hormonally controlled phase, that is, during work continuation and recovery after work (Wolański & Kasprzak 1979; Wolański & Siniarska, 1977).

We have also observed that genetic control of simple traits is stronger compared with more complex traits. Thus, our data suggest that the genetic determination of the 20-m dash time is stronger than that of the 30-m dash time, and that of the 60-m dash time is still weaker. Similar results were obtained using sibling correlations for strength of small muscle groups, that is, stronger genetic control, compared with large muscle groups, weaker genetic control. Static strength has a stronger degree of genetic control compared to explosive power. However, these trends may result from a mathematical regularity in which greater variability coefficients occur for composite measurements (e.g., stature) than for its different components (Wolański, 1962, 1984a).

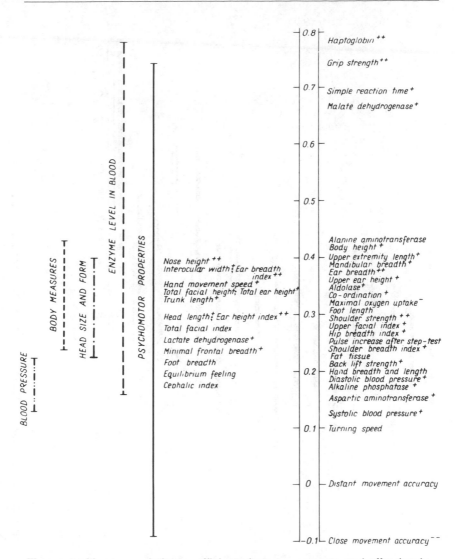

Figure 1. Mean correlation coefficients between parents and offspring in some somatic, physiological, biochemical, and psychomotor traits. Legend: + +positive assortative mating in grandparental and parental generations, +in parental generation only, − −negative assortative mating in grandparental and parental generations, −in parental generation only.

Some data also indicate during which developmental periods various traits are particularly susceptible to training (Wolański, 1979; Wolański & Kasprzak, 1979); these are shown in Figure 2. The role of the different morphological and physiological traits of the motor system in determination of the psychomotor traits, however, is not yet fully elucidated.

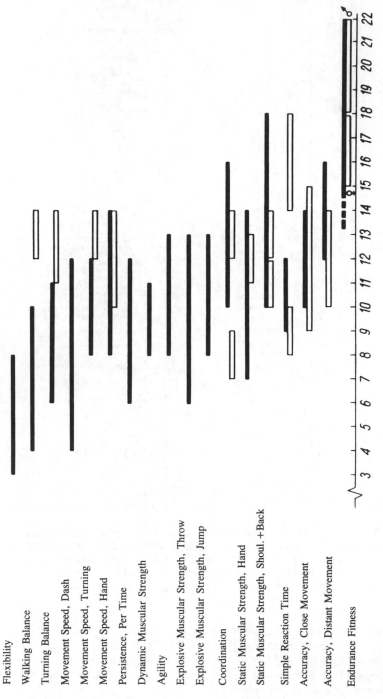

Figure 2. Ages at a high rate of development and low heritability of selected psychomotor traits.

Partitioning Phenotypic Variance

In the widely used model of variation,

$$V_P = V_G + V_N + V_R + e,$$

the sources of variance comprise the unvarying element (V_G—because the genotype does not change during one's lifetime), and the variable element, that is, the environment and mode of life (V_N—the sum of developmental modifiers of nongenetic origin) as well as the genetically determined reactions of the organism to the above modifiers (V_R—normal range of reactions or responses, the interaction and covariance of the unvarying genetic element and variable complex of nongenetic elements). In addition, the element of error is taken into consideration (Wolański, 1984b).

The consequences of this model are very important. Namely, it is thus far assumed that phenotypic variance is the sum of genetic and ecological factors. According to our approach, the greater the participation of variance of genetic origin, the smaller the participation of variance of ecological or nongenetic origin. However, it is not necessarily so since V_G is the measure of genetic polymorphism, which remains unvarying in an individual.

Thus, two phenomena must be separated: the range of genetic variation (scale of polymorphism) in various traits, and the strength of genetic control of trait development. The latter can be, in our opinion, referred to as heritability. According to this view, the susceptiblity of a trait (plasticity) to extra-genetic effects ought to be regarded as reactivity. These two values are complementary and inversely proportional with respect to strength of the effect; that is, when heritability is high, reactivity must be low, and vice versa. The normal range of the reaction modulates the relationships between the strength of genetic control and strength of nongenetic factors. This problem requires separate consideration.

Organism—Environment Relationships

From the standpoint of environmental stimuli, our approach regards the organism as a black box. The environmental stimulus represents the input and the reaction of the organism the response or the output. This model was further developed with the progress of our physiological and ecological studies. Elucidation of the phenomenon of difference in sensitivity to environmental factors (ecosensitivity) suggested that the organism has at its disposal a mechanism decisive of the reception of environmental factors and external defense forces. Due to the latter, individuals (also in various developmental periods) differ in sensitivity to a stimulus identical in strength and duration. The available evidence suggests that the duration of a stimulus, in contrast to its strength, is to a greater extent decisive of the effect assuming that the stimulus does not exceed the strength of the organism. The sensitivity of the organism is genetically determined and modified by experiences early in life (Figure 3). Therefore, apart from strength of the external stimulus, ecosensitivity is decisive of the reception of a stimulus by the organism.

Stimulus reception does not determine the final changes taking place in the organism. The mechanisms regulating the response of the organism depend on the individual's adaptive capacity. It is also under genetic control and depends on earlier experiences, as well as on the unconditioned reflexes specific to the species and the aspirations of representatives of the human species. In the context of this paper, experiences consist of certain behavioral habits and biologically conditioned reflexes. Adaptability decides whether the response of the organism to a stimulus is attenuated or enhanced. In our studies on adaptation of respiratory and circulatory traits to environmental conditions, we observed both overadaptation and nonadaptation, which are manifestations of variation in adaptability (Pyżuk & Wolański, 1972; Wolański & Malik, 1979; Kozioł, 1984).

When so transformed, information and energy are decisive of the kind of response of the organism (Figure 3A, output). This response may involve tem-

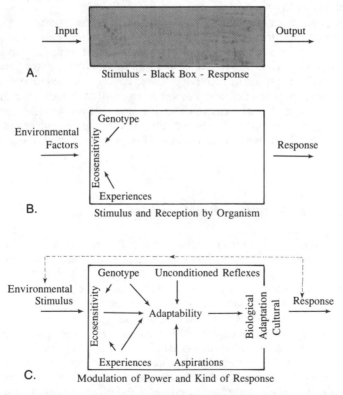

Figure 3. Models of organism-environment relationships: A. older concept of input—black box—output; B. model with ecosensitivity element; C. full model including the reactivity concept (ecosensitivity and adaptability).

porary adaptive changes in the organism itself, that is, reversible, permanent changes, or cultural changes, that is, appropriate behavior, including initiation of changes in the environment itself (Figure 3C).

References

Kozioł, R. (1984). *Respiratory and cardiovascular types of adaptational changes in ontogenetic development in some Polish populations.* Doctoral dissertation, Warsaw.

Pyżuk, M., & Wolańksi, N. (1972). *Respiratory and cardiovascular systems of children under various environmental conditions.* Warsaw: Polish Scientific Publishers.

Siniarska, A. (1979). Parent-child similarity in some cephalometric traits. *Studies in Human Ecology, 3,* 53-70.

Wolański, N. (1962). *Kinetics and dynamics of growth and differentiation in body proportions in children and young people from Warsaw.* Warsaw: Polish Medical Publishers (PZWL).

Wolański, N. (1973a). Concepts and methods in the Polish population studies. *Studies in Human Ecology, 1,* 13-33.

Wolański, N. (1973b). Biologische und soziale Komponenten der motorischen Entwicklung. In K. Willimczik & M. Grosser (Eds.), *Die motorische Entwicklung im Kindes- und Jugendalter* (pp. 324-341). Schorndorf: Karl Hofmann.

Wolański, N. (1984a). Genetics and training possibility of psychomotor traits in man. In N. Wolański & A. Siniarska (Eds.), *Genetics of psychomotor traits in man* (pp. 21-52). Warsaw: International Society of Sport Genetics and Somatology.

Wolański, N. (1984b). Quantitative trait variability of genetic and nongenetic nature. In N. Wolański & A. Siniarska (Eds.), *Genetics of psychomotor traits in man* (pp. 313-323). Warsaw, Poland: International Society of Sport Genetics and Somatology.

Wolański, N., & Kasprzak, E. (1979). Similarity in some physiological, biochemical and psychomotor traits between parents and 2- to 45-year-old offsprings. *Studies in Human Ecology, 3,* 85-131.

Wolański, N., & Malik, S.L. (1979). Modern environment and the future of man. *Acta Anthropogenetica, 3,* 157-162.

Wolański, N., & Pyzuk, M. (1973). Psychomotor properties in 1.5- to 99-year-old inhabitants of Polish rural areas. *Studies in Human Ecology, 1,* 134-162.

Wolański, N., & Siniarska, A. (1977). Genetische bedingungen der menschlichen Entwicklung. In R. Bauss & K. Roth (Eds.), *Motorische Entwicklung* (pp. 275-302). Darmstadt, West Germany: Technische Hochschule.

Wolański, N., Tomonari, K., & Siniarska, A. (1980). Genetics and motor development in man. *Human Ecology and Race Hygiene, 46,* 169-191.

7

Genetics and Cardiac Size

Ted D. Adams, Frank G. Yanowitz, A. Garth Fisher,
J. Douglas Ridges, Arnold G. Nelson, Arthur D. Hagan,
Roger R. Williams, and Steven C. Hunt
THE FITNESS INSTITUTE AT LDS HOSPITAL, SALT LAKE CITY,
UNIVERSITY OF UTAH SCHOOL OF MEDICINE, SALT LAKE CITY,
AND BRIGHAM YOUNG UNIVERSITY, PROVO, UTAH, USA

In the world of sports and athletics, cardiac size and function is considered a major contending factor when the physiological question of what limits human performance is raised. Several echocardiographic studies have been done on various groups of well trained athletes to assess differences in cardiac size and function and how this may relate to the specific mode of training (Allen, Goldberg, Sahn, Schy, & Wojcik, 1977; Gilbert et al., 1977; Morganroth, Maron, Henry, & Epstein, 1975; Roeske, O'Rourke, Klein, & Karliner, 1976; Underwood & Schwade, 1977).

Additional studies have used echocardiography to evaluate cardiac changes that occur among nonathletic individuals who undergo a physical fitness program (Adams et al., 1981; DeMaria, Neumann, Lee, Fowler, & Mason, 1978; Ehsani, Hagberg, & Hickson, 1978; Hanson & Tabakin, 1965). However, very little information has been reported on the influence of heredity on the size, structure, and function of the heart. The question still remains, therefore, as to what extent the variations in cardiac function among individuals engaged in physical performance can be attributed to environmental conditions (primarily exercise training) or genetic endowment.

The purpose of this study was to determine familial (genetic plus family environment) versus nonfamilial influences on cardiac size using data from college-age subjects. Participants included monozygous (MZ) and dizygous (DZ) twins, siblings of like sex and similar in age (SIB), and randomly paired

Part of this study has previously been published in *Circulation*, 17(1):39-49, 1985 and items taken from this publication are done so by permission of the American Heart Association, Inc. The study was supported by funds from the Deseret Foundation and the Research Division of Brigham Young University.

nonrelated subjects (NR). Using twins as subjects, the degree of genetic versus environmental contributions can be estimated since MZ twins have the same genetic makeup and DZ twins are the same as brothers and sisters in their heredity (Klissouras, Pirnay, & Petit, 1973). Echocardiography, electrocardiography, and measurement of maximal oxygen uptake were used to assess intrapair differences between groups before and after exercise training.

Methods

Testing

Forty-one pairs of twins (n = 82: 31 MZ and 10 DZ sets) of both sexes, six pairs of siblings of like sex (SIB) (n = 12: 5 sister pairs and 1 brother pair), and nonrelated male subjects (NR) (n = 30: 15 sets) participated in the study. A computer program randomly paired the nonrelated subjects. All subjects were of college age and the age difference of the SIB pairs was within 12 months. All sets of twins and SIB pairs were raised together, and in most cases both pair members were living together at the time the study was conducted. The twins, SIB pairs, and NR subjects were all in good health and without history of heart disease.

Twins were classified as monozygous or dizygous based on serological analysis. A 10 ml sample of venous blood was drawn from each twin. Seven blood group systems were tested (ABO, Rh, Kell, Duffy, MNS, Lewis, and P), identifying 18 blood group markers. Blood typing was performed using the AABB techniques (American Association of Blood Banks, 1981), and the manufacturer's directions. Discordance in one or more of the blood typings was used as criteria for dizygosity. The twin pairs with identical genotypes were termed syngeneic. Because syngeneicism does not prove monozygosity, the probability of the twins being MZ was calculated (Race & Sanger, 1975) and proved to be greater than 90% in all cases.

Informed consent was obtained from each patient prior to testing. All subjects underwent a health and lifestyle history appraisal, a resting 12-lead electrocardiogram, a maximal graded stress test with measurement of oxygen uptake ($\dot{V}O_2$ max), an M-mode echocardiogram, and a hydrostatic weighing.

The treadmill test protocol called for a constant speed of 6 miles per hour with increasing grades of 2.5% every 2 minutes until the subject reached exhaustion. Subjects were not allowed to use handrails. Expired air was continuously analyzed for percent oxygen and carbon dioxide using an OM-11 02 analyzer and a LB-2 CO2 analyzer (Beckman Instruments). Expired volume was measured using a Parkinson-Cowan high-speed gasometer (Dynascience) and corrected to STPD. An on-line computer (IMSAI 8080) was used to continuously compute oxygen uptake, CO_2 production, and respiratory exchange ratio (R). A five-breath summation was used for analysis. Subjects were hydrostatically weighed using a method previously described (Luft, Carduc, Lim, Anderson, and Howarth (1963).

A standard 12-lead resting electrocardiogram (ECG) was recorded on a Siemens 3170 4-channel instrument. The chest electrodes were placed with great care to ensure consistency of the precordial lead location. ECG recordings were taken at end-tidal volume at paper speeds of 25 and 100 mm/sec for each lead. Measurements of intervals, durations, QRS voltages, and axes were carefully made with calipers. Heart rate and R wave voltages were statistically analyzed for this study.

M-mode echocardiography (ECHO) was performed in a supine and slight right anterior oblique position with a commercially available ultrasonic unit (Primus, Rohe Scientific) using a 3.5 megahertz transducer, long interval focus. All measurements were recorded on an LS-6 fiberoptic recorder (Honeywell) at end-tidal volume using standards developed by the American Society for Echocardiography (Sahn, DeMaria, Kisslo, & Weyman, 1978). A Lead II ECG was also recorded simultaneously. The protocol followed during the ECHO examination has been previously described by Popp, Filly, Brown, and Harrison (1975).

A computerized system was used to interpret the ECHO data. It consisted of a CDC 3300 computer system (Control Data), a BEA-20 CRT terminal; (Beehive International), a model 601 storage monitor (Tektronix), an Omnigraphic Model 6650 X-Y plotter (Houston Instruments), and a Graf/Pen sonic digitizer (Science Accessory). The CRT terminal was used for program control and the storage monitor was used to display graphics as well as alphanumeric messages.

Following the ECHO examination, a well defined segment (at least four cardiac cycles long) of the left ventricular (LV) recording, with the superimposed ECG, was placed on a Graf/Pen digitizer. The technician first entered calibration points from the ECHO recording with regard to time, distance, and placement, as well as the onset of the QRS complex (the fiducial points). ECHO curves and the accompanying ECG signal were then traced by moving the cursor along the leading edge of each relevant ECHO interface. After completing manual tracing, a computer program plotted the traced ECHO data on the X-Y plotter, analyzed the echocardiograms on each successive heart beat, and then averaged the results. The use of this computerized system assured reproducible interpretations, avoided human errors in calculations, and also allowed for continuous measurement and calculation of LV parameters and dimensional changes over several cardiac cycles. This computerized method has previously been used (Adams et al., 1981).

The following echocardiographic measurements were obtained by this computerized system: (a) left ventricular end-diastolic (LVEDD) and end-systolic (LVESD) dimensions, measured at the onset of the QRS complex (F1) and at the point of maximum excursion of the LV posterior wall (F2), respectively; (b) the point of greatest and least dimension between the endocardial surfaces of the interventricular septum and the LV posterobasal wall, measured for LV maximum and minimum dimensions; (c) left ventricular posterobasal wall (LVPW) thickness, measured as the distance between LV endocardium and epicardium at F1 and F2. Based upon the above measurements, the percent shortening of the LV internal diameter was calculated as (LVEDD − LVESD) / (LVEDD x 100).

A method previously described by Teichholz, Kruelen, Herman, & Gorlin (1976) was employed for LV volume calculations. The LV volumes calculated included (a) stroke volume (SV) as the difference between end-diastolic and end-systolic LV volumes; (b) LV wall volume calculated at F1 and F2; and (c) ejection fraction (EF) measured as SV divided by LV volume at end-diastole.

All ECGs were analyzed by one cardiologist and ECHO tracings were made by two echocardiologists. Random numbers were assigned to all subjects' charts to ensure that the cardiologist and echocardiologists would have no knowledge of the status and test sequence of subjects. Seventeen of the ECHO tracings that did not have an adequate number of cycles for computer measurement were measured manually by an echocardiologist.

Training

Following the initial testing component of the study, 14 pairs of MZ twins, 5 pairs of DZ twins, and 6 pairs of SIB members volunteered to participate in 14 weeks of supervised exercise. Both members of the pairs exercised. The training program consisted of 45-minute jogging sessions 4 days per week at 85% maximal measured heart rate. A log was kept of the subject's exercise progress in terms of exercise heart rate, duration, and distance covered during the exercise training sessions. Subjects repeated all tests after successful completion of the 14 weeks of training.

Statistical Analysis

A one-way analysis of variance (ANOVA) was used to test the mean intrapair differences of the four groups (MZ, DZ, SIB, and NR subjects) prior to exercise training. Mean MZ intrapair differences were compared with the combined DZ and SIB mean intrapair differences. This can be justified by the fact that the SIB pairs are as genetically similar as the DZ twins (Klissouras et al., 1973) and their ages are less than 12 months apart. One-way ANOVA was also used to look at postexercise mean intrapair differences for MZ versus combined DZ and SIB pairs. Pre- to postexercise training changes among MZ, DZ, and SIB were also tested. In all cases, *unequal* variances between groups were assumed. Careful review of statistical methodology for genetic research as suggested by Bouchard and Malina (1983a) was made in conjunction with the analysis of the data.

Results

Pretraining

Mean anthropometric, electrocardiographic, and maximal oxygen uptake data and mean intrapair differences for all groups are listed in Table 1. Mean intrapair differences (MID) of MZ twins were significantly less than those of DZ twins and SIB pairs with respect to height and percent body fat ($p < .05$), indicating a strong genetic component for these two variables. Although the mean intrapair weight difference of MZ twins (2.2 kg) was less than for DZ twins (7.3 kg) and SIB sets (6.4 kg), the difference was not significant. The MZ, DZ, and SIB sets had significantly less weight MID than the NR subjects.

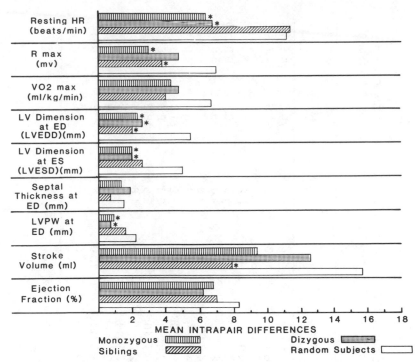

Figure 1. Mean intrapair differences of echocardiograph, electrocardiograph, and VO₂ max data of four groups of college-age subjects: MZ, 31 sets, DZ, 10 sets, SIB, 6 sets, and NR, 15 sets. Mean intrapair differences are represented by bar graphs. R max (mv) = maximal ECG R-wave voltage in millivolts; VO_2 max (ml/kg/min) = maximal oxygen uptake in millimeters of oxygen/kilogram of body weight/minute; LV = left ventricle; EDD = end-diastolic dimension; ESD = end-systolic dimension; PW = posterobasal wall thickness; mm = millimeter; ml = milliliter. * = Significant mean intrapair difference when compared to nonrelated (NR) mean intrapair difference, p < .05.

As indicated in Table 1 and illustrated in Figure 1, there was no significant mean $\dot{V}O_2$ max intrapair difference among the four groups. The ECG data (intrapair difference) showed no significant difference among the MZ, DZ, and SIB groups regarding heart rate and R wave voltages. However, the resting heart rate for MZ (MID = 6.4 beats/min) and DZ (MID = 6.8 beats/min) twins was significantly more similar than for random pairs (MID = 11.1 beats/min). This is graphically illustrated in Figure 1. The ECG voltage, R wave-V5 (also shown in Figure 1), was significantly more similar for MZ, DZ, and SIB pairs compared to NR pairs, while for R wave-max, the MZ and SIB pairs were more similar when compared to NR subjects.

Table 2 lists the mean and mean intrapair difference echocardiographic data of the four groups. Only the left-ventricular posterobasal wall at end-systole

Table 1. Means and standard deviations, and mean intrapair differences (MID) and standard deviations for anthropometric, performance, and electrocardiographic variables in MZ and DZ twin pairs, sibling pairs (SIB), and unrelated pairs (NR) (No. of pairs in parentheses)

Variable	MZ M SD	MZ MID ST	DZ M SD	DZ MID SD	SIB M SD	SIB MID SD	NR M SD	NR MID SD
Weight (kg)	61.8	2.2 (29)*	62.6	7.3 (7)*	62.8	6.4 (6)*	79.0	15.4 (15)
	11.8	2.1	12.4	6.1	10.3	9.3	12.9	11.7
Height (cm)	168.8	1.3 (29)+	168.6	8.2 (7)	168.8	5.7 (6)	—	—
	9.6	1.0	9.3	4.9	8.6	6.2		
Fat %	16.1	1.9 (28)*+	17.8	4.4 (7)*	19.7	4.8 (6)	16.1	9.0 (14)
	6.6	1.3	4.4	2.8	4.6	4.0	6.3	5.6
VO$_2$ max (ml/kg)	51.7	4.3 (29)	49.3	4.8 (7)	44.5	4.0 (6)	50.6	6.7 (15)
	10.7	2.5	7.0	4.7	4.7	3.3	6.4	5.0
Resting Heart Rate	65.3	6.4 (30)*	61.4	6.8 (10)*	59.4	11.2 (6)	61.8	11.1 (14)
	11.0	5.4	8.2	3.8	8.3	6.8	8.8	4.8
R$_{wave}$V5 (mV)	15.5	2.6 (30)*	10.5	4.1 (10)*	12.8	2.8 (6)*	18.3	7.0 (14)
	5.0	2.7	3.7	2.2	4.8	1.6	5.3	3.4
R$_{wave}$V6 (mV)	13.3	2.4 (30)*	9.7	3.1 (10)	11.7	3.0 (6)	14.0	4.7 (14)
	3.9	2.0	3.2	2.5	3.8	1.3	3.6	3.0
R$_{wave}$ max (mV)	16.0	3.0 (30)*	11.7	4.8 (10)	13.0	3.8 (5)*	18.9	7.0 (14)
	5.0	2.7	4.6	3.4	3.5	2.2	5.6	4.4

* = significant intrapair difference when compared to nonrelated subject intrapair difference, p < .05.
+ = significant intrapair difference when compared to combined DZ and SIB, p < .05.

Table 2. Means and standard deviations, and mean intrapair differences (MID) and standard deviations for echocardiographic measurements and volumes in MZ and DZ twin pairs, sibling pairs (SIB), and unrelated pairs (NR). (No. of pairs in parentheses) (LV = left ventricle, ED = end-diastolic, ES = end systolic, and PW = posterobasal wall)

Variable	MZ M SD	MZ MID SD	DZ M SD	DZ MID SD	SIB M SD	SIB MID SD	NR M SD	NR MID SD
LV dimension at ED (LVEDD) (mm) ± SD	44.8 / 5.1	2.3 (26)* / 1.3	44.3 / 4.3	2.6 (10)* / 2.6	46.6 / 4.6	2.0 (6)* / 2.0	46.5 / 4.4	5.5 (14) / 4.3
LV dimension at ES (LVESD) (mm) ± SD	30.4 / 5.4	2.0 (26)* / 1.7	30.9 / 5.0	2.0 (10)* / 2.4	27.6 / 5.8	2.6 (5) / 2.6	33.1 / 4.2	5.0 (14) / 4.5
Septal wall thickness at ED (mm) ± SD	8.8 / 1.6	1.3 (26) / 1.0	9.0 / 1.5	1.2 (9) / 0.9	8.7 / 1.7	0.7 (6) / 0.8	10.7 / 1.7	1.5 (14) / 1.1
Septal wall thickness at ES (MM) ± SD	11.5 / 1.9	1.6 (26) / 1.6	12.5 / 2.2	1.1 (9) / 0.6	11.0 / 2.3	1.2 (5) / 0.7	13.7 / 1.8	1.7 (14) / 1.2
LVPW at ED (mm) ± SD	8.8 / 1.2	0.9 (24)* / 0.7	8.8 / 1.0	0.7 (10)* / 0.6	8.6 / 1.3	1.6 (6) / 0.7	10.6 / 2.1	2.2 (14) / 1.8
LVPW at ES (mm) ± SD	14.8 / 1.7	1.0 (25)* + / 0.9	14.7 / 2.2	1.8 (10)* / 1.2	14.0 / 2.1	2.4 (5) / 1.2	16.0 / 2.4	2.8 (14) / 1.7
LV wall vol. at ED (ml) ± SD	103.7 / 26.7	11.9 (26)* / 12.1	100.7 / 25.1	15.7 (8)* / 9.4	103.6 / 30.3	17.0 (6) / 9.5	136.4 / 27.7	32.8 (14) / 25.5
LV vol. at ED (ml) ± SD	93.2 / 24.2	10.9 (26)* / 6.5	89.4 / 20.9	13.3 (9)* / 10.3	102.2 / 31.0	9.1 (6)* / 8.3	100.8 / 22.3	27.8 (14) / 21.6
LV wall vol. at ES (ml) ± SD	113.8 / 26.8	10.6 (25)* / 8.3	112.6 / 32.1	13.1 (8)* / 9.9	112.2 / 35.9	13.0 (3) / 13.5	145.8 / 25.5	28.4 (14) / 18.5
LV vol. at ES (ml) ± SD	39.2 / 15.2	5.6 (26)* / 4.6	33.8 / 7.8	5.7 (9)* / 6.5	45.4 / 16.6	8.5 (5) / 7.9	45.5 / 13.7	16.4 (14) / 14.3
Stroke vol. (ml) ± SD	53.3 / 13.1	9.4 (26) / 6.6	55.6 / 15.9	12.6 (9) / 10.0	56.2 / 18.6	7.9 (5)* / 2.6	55.2 / 13.2	15.7 (14) / 11.4
Ejection fraction (%) ± SD	58.6 / 8.5	6.8 (26) / 4.5	61.8 / 6.0	6.2 (9) / 7.8	55.3 / 6.9	7.0 (5) / 6.4	55.2 / 7.6	8.3 (14) / 6.2

* = significant intrapair difference when compared to NR, p < .05.

+ = significant intrapair difference when compared to combined DZ and SIB, p < .05.

showed significantly less difference in the MZ twins when compared to the DZ twins and SIB sets combined. As indicated in the table, the MID of the left-ventricular dimension at end-diastole was significantly less ($p < .05$) among MZ (2.3 mm), DZ (2.6 mm), and SIB (2.0 mm) sets when compared to NR pairs (5.5 mm). Intrapair differences of left-ventricular dimension at end-systole was also less among MZ (2.0 mm) and DZ (2.0 mm) ($p < .05$) twins when compared to NR pairs (5.0 mm). SIB (2.6 mm) approached being significantly less than NR ($p < .1$).

There was a trend for MZ, DZ, and SIB pairs to have less difference in stroke volume and ejection fraction than the NR sets, but only the SIB stroke volume MID (7.9 ml) was significantly less than NR (15.7 ml) ($p < .05$). See Figure 1 for a graphical comparison of stroke volume and ejection fraction.

Training

Following the 14 weeks of exercise training, the pooled data from all subjects indicated a significant training effect. The left-ventricular end-diastolic dimension tended to increase although the change did not reach significance (a mean increase of 1.9 mm; $p < .07$ following training). There was a significant decrease in resting heart rate (a mean decrease of 2.6 beats; $p < .01$) and percent body fat (a mean decrease of 2.7%; $p < .01$), and a significant increase in $\dot{V}O_2$ max (a mean increase of 9.2 ml/kg/min; $p < .0001$) and R wave − V5, R wave − V6 (mean increases of 2.3 mV and 2.6 mV, respectively; $p < .05$) after the 14 weeks of exercise.

There were *no* significant changes, however, in mean intrapair differences among MZ versus DZ and SIB sets following training, suggesting that the MZ twin pairs did not become more similar with exercise training.

Discussion

The assessment of functional capacity has wide application in both the clinical testing of patients and the evaluation of human performance. Normal values, expressed as maximal oxygen uptake, $\dot{V}O_2$ max, have been established and correlated with both healthy and diseased populations. Occasionally, however, an individual's $\dot{V}O_2$ max will be measured well above the upper normal range, as is the case with champion endurance runners. The question has been raised as to whether these unusually high values are a result of genetic endowment or environmental conditions (i.e., years of regular exercise training). Using MZ and DZ twins, previous exercise performance studies have been done in which $\dot{V}O_2$ max and other related parameters have been measured (Bouchard et al., 1982; Engström & Fischbein, 1977; Howald, 1976; Klissouras, 1971; Klissouras, Pirnay, & Petit, 1973; Komi & Karlsson, 1979).

Two of the investigations, conducted by Klissouras (Klissouras, 1971; Klissouras et al., 1973), have found the mean intrapair difference of the MZ twins to be significantly less than the intrapair difference of the DZ twins, indicating a strong and predominant genetic contribution. Komi and Karlsson (1979), Howald (1976), and Bouchard et al. (1982), on the other hand, found as much intrapair variation in $\dot{V}O_2$ max among the MZ twins as the DZ twins, suggesting a greater environmental influence. The lack of significant similarity in the maximal oxygen uptake of the MZ twins reported by Komi and Karlsson (1979) was somewhat surprising in that a strong genetic component for slow-twitch muscle fibers was found. Mean intrapair difference of slow-twitch fibers was almost identical in the MZ pairs as compared to a large intrapair difference among the DZ pairs.

Engström and Fischbein (1977), using a large sample size (N = 39 MZ pairs; N = 55 DZ pairs) of boys, found approximately two times as much intrapair difference in maximal oxygen uptake in DZ twins as compared to MZ twins. When the amount of physical exercise performed during leisure time activity was controlled for, however, the MZ pairs were as dissimilar as the DZ pairs. Engtröm and Fischbein suggested that the phenotype, physical work capacity, may be strongly influenced by the environment, which would include physical training. Our findings of no significant $\dot{V}O_2$ max MID among the four groups support those of Komi and Karlsson (1979), Howald (1976), and Bouchard et al. (1982), suggesting that genetic factors alone are not as significant in establishing maximal oxygen uptake as has been previously thought.

Since $\dot{V}O_2$ max is the product of the cardiovascular delivery system and the peripheral oxygen extraction, this study attempted to ascertain what contribution the blood transport system played in the degree of intrapair variation. Both Klissouras (1971) and Howald (1976), prior to the refinement of echocardiography, approximated heart volumes from chest x-rays and found that MZ twins demonstrated as much mean intrapair variability as DZ twins. In a more recent echocardiographic study, Diano et al. (1980) reported a significant familial similarity in left ventricular dimensions in members of nuclear families. A study investigating the genetic influence upon heart rate and the electrocardiogram has recently been reported by Havlik, Garrison, Fabsitz, and Feinleib (1980). They reported a modest genetic influence (35%) upon the P-R duration and a 54% variance in heart rate due to genetic variation. No genetic influence was reported on the variability of QRS duration and QT interval, nor were data reported regarding R wave voltages. Our study showed the greatest influence on cardiac size to be familial.

It was interesting to note the similarity of the ECG waveforms of some of the MZ pairs. Figure 2 illustrates the ECG waveform resemblance found in one set of 23-year-old MZ twins. To further investigate the electrocardiographic similarities of the twins, siblings, and nonrelated subjects, our laboratory is now in the process of analyzing the following additional ECG parameters: PR interval, QRS interval, QT interval, P axis, QRS axis, T axis, RR interval, and frontal and horizontal plane voltages.

As indicated in the methods section, both members of the twin sets and SIB pairs were asked to train for a 14-week period. If with exercise training the MZ twins obtained identical gains in cardiac size and $\dot{V}O_2$ max and/or reached an identical potential while the nonidentical twins and SIB pairs had unequal cardiac and $\dot{V}O_2$ training gains, then the role of genetic influence could be considered stronger than the environmental factors. Following exercise training, however, the MZ twins still showed as much intrapair difference as DZ twins and SIB pairs with regards to cardiac size and $\dot{V}O_2$ max.

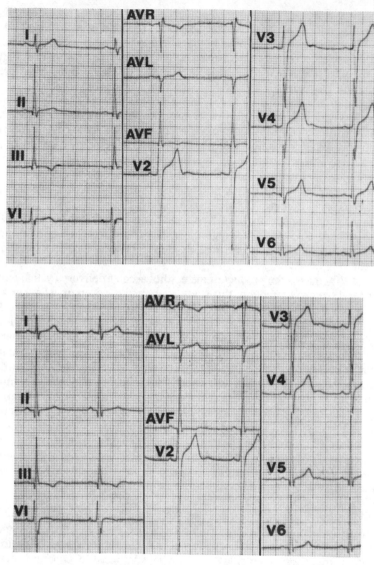

Figure 2. Electrocardiograms of a 23-year-old set of male MZ twins. Twin A is at the top of the figure; twin B is at the bottom.

Bouchard and Malina (1983), in a recent review article, indicated there is evidence of a genotype-environment interaction regarding the training of maximal aerobic power. The question still remains as to whether a 14-week training period is long enough for the possible genotype-environment interaction to fully express itself. Landry et al. (1983) recently reported on a small population of MZ twins (9 sets) who underwent 20 weeks of endurance training. Echocardiographic data were obtained before and after exercise training. The posttraining intrapair difference was more similar when compared to pretesting, suggesting a possible "homogenizing effect" of the exercise training on cardiac size and structure. Further twin exercise training studies with larger sample sizes are needed to verify these findings and the ones found in our study.

In view of the genotype-environment interaction discussed by Bouchard and Malina (1983a) and the recent echocardiograph study reported by Landry et al. (1983), reference can be made to a set of MZ twins participating in our study. The MZ pair, 24 years of age, had been raised together and had equal exercise histories. Both twins trained regularly, each was running 30 to 45 minutes per day as well as participating in intramural sports. Their echocardiographic data, shown in Figure 3, shows very little intrapair difference in left ventricular dimensions, posterobasal wall thickness, and septal wall thickness.

In conclusion, the greater similarity in cardiac size between MZ twins, DZ twins, and SIB sets as compared to the NR sets suggests that familial influences, which include common environment plus genetic factors, are more important determinants of cardiac size than are nonfamilial or purely genetic influences. Further studies are needed to fully characterize the specific contributions made by environmental and/or genetic factors and how each of these factors interact to produce a training effect on cardiovascular parameters.

Addendum

To further evaluate the findings of this study, our laboratory is currently testing a second cohort of twins in which an equal number of MZ and DZ twins are participating. These twin sets range in age from 19 to 24 years and include both male and female sets. The ECHO and functional capacity methodology used to test these twins has been altered somewhat, and additional cardiopulmonary measurements have been included.

Both M-mode and two-dimensional echocardiography are being used to test these subjects. The two-dimensional ECHO ensures greater accuracy in locating the various anatomical landmarks of the heart for M-mode measurement. In addition to the left-ventricular parameters that were measured and calculated in the previous study, our laboratory has also been analyzing the left atrium, the right heart, aortic diameter, and cardiac measurement ratios. Besides the 12-lead electrocardiographs, each twin is undergoing a 32-lead electrocardiographic mapping to analyze ECG depolarization/repolarization patterns over the body surface. ECG voltages are entered into a computer (on-line), and isopotential maps for various time sequences are produced.

Blood pressure is being measured in a variety of clinical settings. These include measurements taken during isometric grip on a tilt table at a 50-degree tilt, in a sitting or standing or supine resting position, and during exercise at various work rates. To further investigate possible blood pressure related parameters, measurement of ATP-dependent sodium/potassium transport is being analyzed using plasma blood samples. Additionally, ouabain insensitive cation transport measurements (including sodium/lithium countertransport and lithium/potassium cotransport) are also being taken.

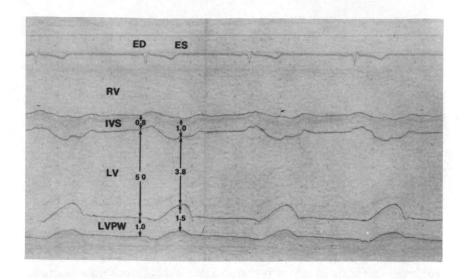

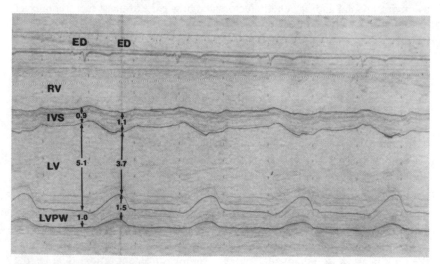

Figure 3. Echocardiograms of a 24-year-old set of male MZ twins. Twin A is at the top of the figure; twin B is at the bottom.

To evaluate the maximal work performance of the twins, our laboratory chose to work the participants on an electronically braked bicycle ergometer (Siemens-Elema 320B). The ergometer is controlled via a 12-bit digital-to-analog computer. During the exercise test, the gas response data of the subject is measured by a computerized on-line breath-by-breath stress test system. Breath-by-breath and four-breath average data is computed and stored on disk, while averaged data for every four breaths is printed during the exercise test. A sample plot of the four-breath averaged data is shown in Figure 4. This system has previously been described (Yeh, Gardner, Adams, Yanowitz, & Crapo, 1983).

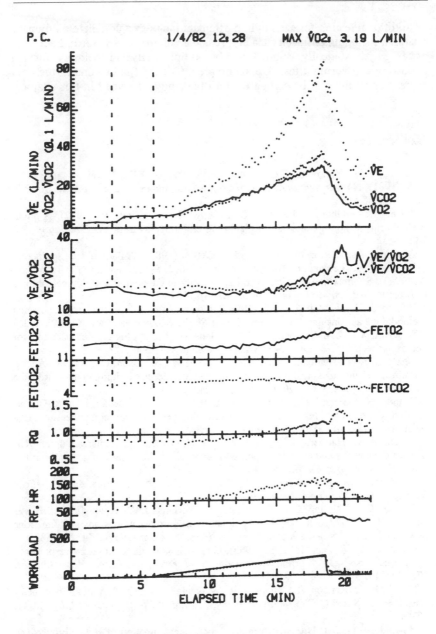

Figure 4. Sample plot of gas response data. The 6 frames, from top to bottom: (1) minute ventilation (VE, 1/min), O_2 consumption rate (VO_2, 0.1 1/min), and CO_2 production rate (VCO_2, 0.1 1/min); (2) ventilatory equivalent for O_2 (VE°VO_2) and ventilatory equivalent for CO_2 (VE/VCO2); (3) fractional end-tidal (O_2, in %) and fractional end-tidal CO_2 (FETCO_2, in %); (4) respiratory quotient (RQ); (5) heart rate (HR) and respiratory frequency (RF); and (6) work load (in W).

Additional testing for each twin participant includes a complete pulmonary analysis, including diffusion capacity, residual volume, and so forth. Each twin is being hydrostatically weighed, and blood lipid analysis is being performed following an overnight fast. Upon completion of the final testing of the twin pairs, the results will be compared with the study described in this chapter.

References

Adams, T.D., Yanowitz, F.G., Fisher, A.G., Ridges, J.D., Lovell, K., & Pryor, T.A. (1981). Noninvasive evaluation of exercise training in college-age men. *Circulation*, **64**, 958-965.

Allen, D., Goldberg, S.J., Sahn, D.J., Schy, N., & Wojcik, R. (1977). A quantitative echocardiographic study of champion childhood swimmers. *Circulation*, **55**, 142-145.

American Association of Blood Banks (AABB) (1981). Philadelphia: Lippincott.

Bouchard, C., Leblanc, C., Lortie, G., Simoneau, J.A., Theriault, G., & Tremblay, A. (1982). Submaximal physical working capacity in adopted and biological siblings (abstract). *Medicine and Science in Sports*, **14**, 139.

Bouchard, C., & Malina, R.M. (1983a). Genetics for the sport scientist: Selected methodological consideration. *Exercise and Sports Science Reviews*, **11**, 275-303.

Bouchard, C., & Malina, R.M. (1983b). Genetics of physiological fitness and motor performance. *Exercise and Sport Science Reviews*, **11**, 306-339.

DeMaria, A.N., Neumann, A., Lee, G., Fowler, W., & Mason, D.T. (1978). Alterations in ventricular mass and performance induced by exercise training in man evaluated by echocardiography. *Circulation*, **57**, 237-243.

Diano, R., Bouchard, C., Dumesnil, J., LeBlanc, C., Laurenceau, J.L., & Turcot, J. (1980). Parent-child resemblance in left ventricular echocardiographic measurements (abstract). *Canadian Journal of Applied Sports Science*, **5**, 4.

Ehsani, A.A., Hagberg, J.M., & Hickson, R.C. (1978). Rapid changes in left ventricular dimensions and mass in response to physical conditioning and deconditioning. *American Journal of Cardiology*, **42**, 52-56.

Engström, L.M., & Fischbein, S. (1977). Physical capacity in twins. *Acta Geneticae Medica et Gemellologiae*, **26**, 159-165.

Gilbert, C.A., Nutter, D.O., Felner, J.M., Perkins, J.V., Heymsfield, S.B., & Schlant, R.C. (1977). Echocardiographic study of cardiac dimensions and functions in the endurance-trained athlete. *American Journal of Cardiology*, **40**, 528-533.

Hanson, J.S., & Tabakin, B.S. (1965). Comparison of the circulatory response to upright exercise in 25 "normal" men and 9 distance runners. *British Heart Journal*, **27**, 211-219.

Havlik, R.J., Garrison, R.J., Fabsitz, R., & Feinleib, M. (1980). Variability of heart rate, P-R, QRS and Q-T durations in twins. *Journal of Electrocardiography*, **13**(1), 45-48.

Howald, H. (1976). Ultrastructure and biochemical function of skeletal muscle in twins. *Annals of Human Biology*, **3**(5), 455-462.

Klissouras, V. (1971). Heritability of adaptive variation. *Journal of Applied Physiology*, **31**(3), 338-343.

Klissouras, V., Pirnay, F., & Petit, J.M. (1973). Adaptation to maximal effort: Genetics and age. *Journal of Applied Physiology*, **35**(2), 288-293.

Komi, P.V., & Karlsson, J. (1979). Physical performance, skeletal muscle, enzyme activities, and fiber types in monozygous and dizygous twins of both sexes. *Acta Physiologica Scandinavica* (Supplement), **462**, 1-28.

Landry, F., Bouchard, C., Prud'homme, D., Diano, R., Leblanc, C., & D'Amours, Y. (1983). Cardiac dimensions in twins following endurance training (abstract). *Medicine and Science in Sports and Exercise*, **15**(2), 125.

Luft, U.C., Carduc, D., Lim, T.P., Anderson, E.C., & Howarth, J.L. (1963). Physical performance in relation to body size and composition. *Annals of the New York Academy of Science*, **110**, 795-808.

Morganroth, J., Maron, B.J., Henry, W.L., & Epstein, S.E. (1975). Comparative left ventricular dimensions in trained athletes. *Annals of Internal Medicine*, **82**, 521-524.

Popp, R.L., Filly, K., Brown, O.R., & Harrison, D.C. (1975). Effect of transducer placement on echocardiographic measurement of left ventricular dimensions. *American Journal of Cardiology*, **35**, 337-540.

Race, R.R., & Sanger, R. (1975). *Blood groups in man* (6th ed.). London: Blackwell Scientific.

Roeske, W.R., O'Rourke, R.A., Klein, A.L., & Karliner, J.S. (1976). Noninvasive evaluation of ventricular hypertrophy in professional athletes. *Circulation*, **53**, 286-291.

Sahn, D.J., DeMaria, A., Kisslo, J., & Weyman, A. (1978). Recommendations regarding quantitation in M-mode echocardiography: Results of a survey of echocardiographic measurements. *Circulation*, **58**, 1072-1082.

Teichholz, L.E., Kruelen, T., Herman, M.V., & Gorlin, R. (1976). Problems in echocardiographic volume determinations: Echocardiographic-angiographic correlations in the presence or absence of asynergy. *Annals of the Journal of Cardiology*, **37**, 7-11.

Underwood, R.H., & Schwade, J.L. (1977). Noninvasive analysis of cardiac function of elite distance runners-Echocardiography, vectocardiography, and cardiac intervals. *Annals of the New York Academy of Science*, **301**, 297-309.

Yeh, P., Gardner, R., Adams, T., Yanowitz, F., & Crapo, R. (1983). "Anaerobic threshold": Problems of determination and validation. *Journal of Applied Physiology*, **55**(4), 1178-1186.

8

Muscle Fiber Type Composition and Enzyme Activities in Brothers and Monozygotic Twins

Gilles Lortie, Jean-Aimé Simoneau, Marcel R. Boulay, and Claude Bouchard
LAVAL UNIVERSITY
QUEBEC, CANADA

Well-trained endurance athletes have a higher VO2 max than untrained subjects. In addition to high VO2 max and endurance capability, endurance athletes also exhibit a higher percentage of type I muscle fibers and muscle oxidative enzyme activities than untrained or moderately active individuals (Howald, 1982). Studies on the effects of endurance training have indicated that the activity of oxidative enzymes may be increased by 100% (Gollnick, Armstrong, Saltin, Saubert, & Shepherd, 1973). Endurance training also induces a progressive shift from type IIb to type IIa fibers in skeletal muscle (Andersen & Henriksson, 1977), while no evidence of change in the percentage of type I fibers has yet been reported in humans.

Several authors have thus concluded that the absence of change in the percentage of type I fibers with training was due to the fact that fiber composition of skeletal muscle was under a strict genetic control. This in turn implies that it was determined early in life and remained constant thereafter. To reinforce this point, the study by Komi, Viitasalo, Havu, Thorstensson, Sjodin, and Karlsson (1977), which reported heritability coefficients of 0.93 or higher in a small sample of DZ and MZ twins in both sexes, is frequently quoted.

Thanks are expressed to P. Hamel, C. Leblanc, L. Perusse, R. Savard, A.M. Simoneau, and G. Theriault for their assistance in the course of this study. This research was supported by NSERC of Canada (A-8150).

However, twin studies with cross-sectional observations are not always easy to interpret and the data are often affected by methodological problems (see Bouchard & Malina, 1983). The aim of this study was to provide additional data on sibling resemblance in skeletal muscle fiber type distribution and area, and in enzyme activities.

Methods

Thirty-two pairs of brothers from 20 sibships (age 25 ± 4 years, weight 69 ± 10 kg, height 172 ± 6 cm), and 35 pairs of male and female MZ twins (age 21 ± 3 years, weight 62 ± 10 kg, height 167 ± 9 cm), gave written consent to participate in this study and were advised of the risk and discomfort associated with the muscle biopsy. All the procedures were approved by the Medical Ethics Committee of Laval University.

Muscle biopsy: Muscle biopsy was obtained from the middle of the vastus lateralis, 12 to 16 cm above the patella and approximately 2 cm away from the epimysium by the percutaneous needle biopsy technique modified by Evans, Phinney, and Young (1982).

Histochemical analysis: Based on staining properties for myofibrillar ATPase, the different fibers were designated as type I (i.e., not stained), type IIa (lightly stained) and type IIb (darkly stained) on the same section according to the technique described by Mabuchi and Sreter (1980). For each specimen, at least 300 fibers were classified by two different evaluators and the mean value was retained. The mean muscle fiber area was determined by averaging the cross-sectional areas of 20 randomly selected fibers of each type.

Biochemical analysis: The activities of hexokinase (HK), phosphofructokinase (PFK), lactate dehydrogenase (LDH), malate dehydrogenase (MDH), 3-hydroxyacyl CoA dehydrogenase (HADH), and oxoglutarate dehydrogenase (OGDH) were assayed using pyridine nucleotide enzyme reactions according to the principles outlined by Lowry and Passonneau (1972). All procedures for determining fiber type distribution, area, and enzyme activities have been previously described in more detail by Simoneau et al. (1985).

Statistical procedures: Simple and multiple regression analyses were performed in order to adjust for age differences in brothers, and for age and sex differences in MZ twins for all measurements. Residual scores were computed and submitted to analysis of variance and simple interclass correlation. The analysis of variance and F-ratio computed from the between-sibships over the within-sibship variance were obtained following the procedures outlined in Haggard (1958).

Results

Descriptive statistics for fiber type composition, fiber area, and enzyme activities in brother and MZ twin sibships are presented in Table 1, while correlations between muscle characteristics and age in brothers, and age and sex in MZ twins, are presented in Table 2. The effect of age ranged from 0% to 7% in brothers, and the combined effects of age and sex varied between 1 and 46% in male and female MZ twins.

All muscle characteristic data were adjusted for the effect of age in brothers, and the effect of age and sex in MZ twins by simple and multiple regression analyses, respectively. The adjusted score was then the residual between the original and the predicted score. These adjusted scores were submitted to interclass correlation analysis and analysis of variance in order to test for the presence of sibling resemblance in muscle characteristics. These results are shown in Table 3. MZ twins exhibited significant similarity for the percentage of type I and type IIb muscle fibers, and for all enzyme activities, while only the percentage of type I fibers, LDH, and HADH had significant resemblance within the brother sibships. In addition, the ratio of PFK to OGDH was significant in both the biological brothers and MZ twins.

Discussion

Fiber Type Composition

Results of this study indicate the presence of significant biological resemblance in the percentage of type I fibers in MZ twins and in brothers. Komi et al. (1977) reported that MZ twins (n = 15 pairs) were almost identical in the percentage of type I fibers in the vastus lateralis, while DZ twins (n = 16 pairs) were quite variable. Heritability coefficients reached 0.93 and above,

Table 1. Means and standard deviations of fiber and enzyme variables for the study in brothers and monozygotic (MZ) twins

Variable	Brothers $n = 46$	MZ twins $n = 70$
% Type I	43 ± 12[a]	51 ± 14
% Type IIa	36 ± 12	34 ± 11
% Type IIb	21 ± 11	15 ± 9
Type I area (μm^2)	5173 ± 1097	4114 ± 1010
Type IIa area (μm^2)	5466 ± 1075	4197 ± 1208
Type IIb area (μm^2)	4768 ± 1520	3912 ± 1376
HK[b]	1.1 ± 0.3	1.2 ± 0.3
PFK[b]	80 ± 33	134 ± 32
LDH[b]	187 ± 84	206 ± 99
MDH[b]	175 ± 54	212 ± 59
HADH[b]	5.2 ± 2.1	3.9 ± 1.3
OGDH[b]	0.83 ± 0.44	0.62 ± 0.26
PFK/OGDH	124 ± 83	269 ± 201

Note. HK = hexokinase; PFK = phosphofructokinase; LDH = lactate dehydrogenase; MDH = malate dehydrogenase; HADH = hydroxyacyl CoA dehydrogenase; and OGDH = oxoglutarate dehydrogenase.

[a]$M \pm SD$.
[b]Activity in μmol NADH (NADPH)/g of wet tissue · min^{-1}·

Table 2. Simple and multiple correlations of age and sex on muscle characteristics in brothers and monozygotic (MZ) twins

Variable	Brothers (n = 46) age r	MZ twins (n = 70) age r	sex r	R^a
% Type I	0.14	-0.11	0.27*	0.31*
% Type IIa	-0.22	0.20	-0.22	0.32*
% Type IIb	0.09	-0.04	-0.11	0.11
Type I area	0.26	0.00	-0.34**	0.34*
Type IIa area	0.22	-0.07	-0.60**	0.60**
Type IIb area	0.16	-0.05	-0.68**	0.69**
HK	-0.16	-0.10	-0.20	0.21
PFK	-0.03	-0.06	-0.27*	0.27
LDH	-0.26	0.05	-0.52**	0.53**
MDH	0.09	-0.16	-0.27*	0.29
HADH	0.11	-0.26	-0.15	0.28
OGDH	0.18	0.04	0.05	0.06
PFK/OGDH	0.02	-0.03	-0.11	0.11

Note. HK = hexokinase; PFK = phosphofructokinase; LDH = lactate dehydrogenase; MDH = malate dehydrogenase; HADH = hydroxyacyl CoA dehydrogenase; and OGDH = oxoglutarate dehydrogenase.
[a]From Y = age + sex.
*p ≤ 0.05, **p ≤ 0.01.

Table 3. Interclass and intraclass correlations in pairs of brothers and monozygotic (MZ) twins for fiber type distribution, fiber areas, and enzyme activities

Variable	Brothers[a] (n = 32 pairs) Interclass correlation	Intraclass coefficient	F ratio	MZ twins[b] (n = 35 pairs) Interclass correlation	Intraclass coefficient	F ratio
% Type I	0.37*	0.33	2.2*	0.54	0.55	3.4**
% Type IIa	-0.04	-0.03	0.9	0.17	0.18	1.4
% Type IIb	0.24	0.26	1.8	0.59**	0.56	3.6**
Type I area	0.34	0.29	1.9	0.33	0.30	1.9*
Type IIa area	0.31	0.30	2.0	-0.08	-0.08	0.8
Type IIb area	0.22	0.20	1.6	0.03	0.05	1.0
HK	-0.28	-0.22	0.6	0.42**	0.41	2.4**
PFK	0.27	0.27	1.8	0.55**	0.55	3.4**
LDH	0.51	0.50	3.3**	0.72**	0.68	5.2**
MDH	0.13	0.15	1.4	0.58**	0.58	3.8**
HADH	0.45**	0.48	3.1**	0.44**	0.43	2.5**
OGDH	0.10	0.09	1.2	0.56**	0.53	3.2**
PFK/OGDH	0.51*	0.34	2.1*	0.35**	0.30	1.9**

Note. HK = hexokinase; PFK = phosphofructokinase; LDH = lactate dehydrogenase; MDH = malate dehydrogenase; HADH = hydroxyacyl CoA dehydrogenase; and OGDH = oxoglutarate dehydrogenase.
[a]Scores adjusted for age.
[b]Scores adjusted for age and sex.
*p ≤ 0.05, **p ≤ 0.01.

and they postulated that muscle-specific fiber composition was primarily genetically determined with little or no significant nongenetic influences. However, other data suggest that this coefficient (0.93) is overestimated.

Unpublished data from our laboratory showed that the reliability coefficient of the percentage of type I fiber determination was 0.88, suggesting that in our laboratory approximately 12% of the total interindividual variation in the percentage of type I fibers was due to experimental error. MZ twins of the present study were characterized by an intraclass coefficient of only 0.55, indicating that other factors must also contribute to the total variance. In fact, the difference between the remaining variance (45%) and measurement error (12%) indicates that about 30% of the total variance can be accounted for by nongenetic factors. Moreover, other more recent studies suggest that nongenetic influences contribute to the relative distribution of type I fibers. Thus, both Jansson, Sjodin, and Tesch (1977) and Schantz, Billeter, Henriksson, and Jansson (1982) have proposed that some transformation between type II and type I fibers may occur, and that the intermediate fiber type, frequently designated as "type IIc," constitutes a transitional phase of this process.

In addition, Monster, Chan, and O'Connor (1978) have suggested that increased use of a muscle eventually leads to a rise in the percentage of type I fibers and a reduction in speed of contraction. Fugl-Meyer, Eriksson, Sjostrom, and Soderstrom (1982) have noted that functional demands are also important determining factors in the development of muscle structure. Their observations on right/left muscle morphological asymmetry were attributed to different right/left functional demands, leading to some degree of functional adaptation in fiber composition. These observations would thus seem to suggest that heritability coefficient for the percentage of type I muscle fibers is probably much lower than previously reported.

No study to date has presented data on sibling resemblance in classes of type II fibers. In the present study, there was a significant resemblance only for the percentage of type IIb fibers in MZ twins. The intraclass coefficient of 0.59 must be viewed with caution, however, as results from our laboratory suggest that determination of the relative distribution of type IIb fibers is only about 60% reliable. In other words, the correlation between MZ twins may be associated with a true genetic covariation effect, a similar life and physical activity pattern, or measurement error. It should also be noted that Andersen and Henriksson (1977) have shown that endurance training can induce a significant shift from type IIb to type IIa, and this progressive transformation may lead to the absence of type IIb fibers often observed in well-trained endurance athletes.

Fiber Areas

There was no significant resemblance in brothers for any of the fiber areas, while MZ twins exhibited a significant resemblance for only type I fiber area. However, coefficients were quite low in both types of sibships, suggesting that genetic factors are probably only slightly involved in the interindividual variation of fiber type areas. In fact, studies of endurance or strength training have shown that fiber type areas are enhanced as a function of the duration and intensity of training (Howald, 1982). To our knowledge, no studies have presented sibship resemblance data for fiber type areas. Nevertheless, in analyses of skeletal muscle ultrastructure of 11 MZ and 6 DZ twin pairs, Howald (1976) reported no gene-associated variation in the mitochondrial density, the ratio of mitochondrial volume to myofibril volume, or the internal and external surface densities of mitochondria. It was concluded that ultrastructural features of skeletal muscle were more dependent upon environmental influences than on genetic factors.

Enzyme Activities

Resemblances in enzyme activities were systematically significant in MZ twins but less consistently so in brothers. When comparisons were made between both types of siblings, only PFK (although nonsignificant in brothers) and the PFK-to-OGDH ratio provided some indication of a moderate genetic effect. It has recently been suggested that both PFK and OGDH are important enzymes in their respective metabolic pathways (Newsholme, 1980). In the present study these two enzyme activities did not present any significant resemblance in brothers, but the ratio of PFK to OGDH did ($r = 0.41$). MZ twins were also characterized by a significant resemblance for this ratio. The data thus suggest that genetic factors may be involved in variation of the ratio of aerobic to glycolytic activity, but not in variation of enzymes considered individually and outside their full metabolic path.

In the literature, Howald (1976) showed that muscle HK and succinate dehydrogenase (SDH) activities were identical in both DZ and MZ twins, while muscle glyceraldehyde 3-phosphate dehydrogenase (GPDH) and HADH activities were significantly more variable within DZ than MZ twin pairs. Even though this was indicative of a genotypic influence, there were significant mean differences between both sets of twins for GPDH and for HADH, and a significant difference in variance for GPDH between twin types. As such differences are known to introduce bias in the analysis of twin data (Christian, 1979), one should view these results with caution. On the other hand, in the study reported by Komi et al. (1977), there was no evidence of significant genetic variation in activities of several skeletal muscle enzymes from DZ and MZ comparisons; nevertheless, the results must be considered with caution.

In summary, results of the present study indicate a significant resemblance for the percentage of type I fibers in MZ twins and in brother sibships. The coefficients obtained suggest only a moderate hereditary effect on fiber type distribution, however. The low coefficients found in both types of siblings suggest that variation in fiber type areas is primarily related to nongenetic influences. Finally, although there was significant resemblance in MZ twins for the enzyme activities, the low and often nonsignificant intraclass coefficients generally observed in brothers suggest that variation in enzyme activities alone is probably more related to environmental and nongenetic factors than to inheritance. The ratio of aerobic to glycolytic enzyme activities, however, might be moderately associated with genetic factors.

References

Andersen, P., & Henriksson, J. (1977). Capillary supply of the quadriceps femoris muscle of man: Adaptive response to exercise. *Journal of Applied Physiology*, **270**, 277-690.

Bouchard, C., & Malina, R.M. (1983). Genetics for the sport scientist: Selected methodological considerations. *Exercise and Sport Sciences Reviews*, **11**, 275-305.

Christian, J.C. (1979). Testing twin means and estimating genetic variance. Basic methodology for analysis of quantitative twin data. *Acta Geneticae Medicae et Gemellologiae*, **28**, 35-40.

Evans, W.J., Phinney, S.D., & Young, V.R. (1982). Suction applied to a muscle biopsy maximizes sample size. *Medicine and Science in Sports and Exercise*, **14**, 101-102.

Fugl-Meyer, A.R., Eriksson, A., Sjostrom, M., & Soderstrom, G. (1982). Is muscle structure influenced by genetical or functional factors? A study of three forearm muscles. *Acta Physiologica Scandinavica*, **114**, 277-281.

Gollnick, P.D., Armstrong, R.B., Saltin, B., Saubert C.W. IV, & Shepherd, R.R. (1973). Effect of training on enzyme activity and fiber composition of human skeletal muscle. *Journal of Applied Physiology*, **34**, 107-111.

Haggard, E.A. (1958). *Intra-class correlation and the analysis of variance*. New York: Dryden Press.

Howald, H. (1976). Ultrastructure and biochemical functions of skeletal muscle in twins. *Annals of Human Biology*, **3**, 455-462.

Howald, H. (1982). Training-induced morphological and functional changes in skeletal muscle. *International Journal of Sports Medicine*, **3**, 1-12.

Jansson, E., Sjodin, B., & Tesch, P. (1977). Changes in muscle fiber distribution in man after physical training. A sign of fiber type transformation? *Acta Physiologica Scandinavica*, **104**, 235-237.

Komi, P.V., Viitasalo, J.H.T., Havu, M., Thorstensson, A., Sjodin, B., & Karlsson, J. (1977). Skeletal muscle fibers and muscle enzyme activities in monozygous and dizygous twins of both sexes. *Acta Physiologica Scandinavica*, **100**, 385-392.

Lowry, O.H., & Passonneau, J.V. (1972). *A flexible system of enzymatic analysis*. New York: Academic Press.

Mabuchi, K., & Sreter, F.A. (1980). Actomyosin ATPase. II. Fiber typing by histochemical ATPase reaction. *Muscle and Nerve*, **3**, 233-239.

Monster, A.W., Chan, H.C., & O'Connor, D. (1978). Activity patterns of human skeletal muscles: Relation to muscle fiber type composition. *Science*, **200**, 314-317.

Newsholme, E.A. (1980). Use of enzyme activity measurement in studies of the biochemistry of exercise. *International Journal of Sports Medicine*, **1**, 100-102.

Schantz, P., Billeter, R., Henriksson, J., & Jansson, E. (1982). Training-induced increase in myofibrillar ATPase intermediate fibers in human skeletal muscle. *Muscle and Nerve*, **5**, 628-636.

Simoneau, J.A., Lortie, G., Boulay, M.R., Thibault, M.C., Theriault, G., & Bouchard, C. (1984). Skeletal muscle histochemical and biochemical characteristics in sedentary male and female subjects. *Canadian Journal of Physiology and Pharmacology*, **60**, 30-35.

9

Genetic Variation in the Force-Velocity Relation of Human Muscle

Brian Jones
MCGILL UNIVERSITY
MONTREAL, CANADA

Vassilis Klissouras
UNIVERSITY OF ATHENS
ATHENS, GREECE

Individual differences observed in maximal performance of activities such as sprinting, throwing, and jumping may be attributed primarily to existing differences in maximal muscular power (MMP) provided by the contracting muscles. The MMP determined from measurements of the maximal running velocity on a staircase and the body weight shows wide variability which seems to be the result of genetic diversity (Komi, Klissouras, & Karvinen, 1973). Since the MMP depends, among others, on the relative position of the force-velocity curve, which is a function of the maximal velocity of shortening (Vo) and the maximal isometric force (Po), we have extended our previous work by ascertaining to what extent heredity contributes to inter-individual variation in the force-velocity relation and its two components, namely, Vo and Po.

Methods

Twins

Nine pairs of male monozygotic twins and eight pairs of like-sexed dizygotic twins, who lived at home with their parents, were used as subjects. They ranged in age from 11 to 17 years. Zygosity was determined on the basis of morphological traits and serological examination in a manner previously described (Klissouras, 1971).

The Ergometer

An ergometer was specifically designed for this study to permit determination of the force-velocity relationship of the forearm flexor muscles as they collectively exerted their force at the palm of the hand.[1] The ergometer was constructed in such a way that minor adjustments allowed the right or left arm to be tested. A conceptual arrangement of the ergometer is shown in Figure 1.

The subject was seated and placed his dominant upper arm on a rest at a right angle to his upper body and in line with his shoulders. Adjusting the height of the armrest or the seat helped maintain this position. Further, the elbow joint was kept directly in line with the pivot axis of the lever arm of the ergometer by adjusting side pads and a securely fastened safety belt. The subject gripped an adjustable handle and moved the lever arm of the ergometer with a maximal voluntary contraction. For all dynamic contractions in the present investigation, the lever arm of the ergometer, and thus the forearm of the subject, was set at an experimental starting position of 60° with respect to the horizontal. The entire range of movement was 50°.

A potentiometer was attached directly to the pivot axis of the lever arm of the ergometer and coincided with the axis of rotation of the subject's elbow joint. The potentiometer was connected to a Honeywell Model 1706 Visicorder Oscillograph. Therefore, throughout the movement the instantaneous angular displacement of the forearm as a function of time was recorded.

Force was exerted against a weight placed at a variable distance from the pivot axis of the resistance arm of the ergometer, which was at a right angle to the lever arm. The lever and resistance arm of the ergometer was statically balanced with counterweights so that the subject worked against only the moment about the pivot axis generated by the weight plus the inertial forces of the frame structure and human forearm.

With the same weight, a series of different loads was obtained by manually moving the weight down the shaft of the resistance arm away from the pivot axis of rotation. The distance which the weight was moved was accurately determined by a numerical counter and corresponded to the number of revolutions taken by the manual crank. Therefore, the exact load could easily be reproduced between twin pairs.

Maximal Linear Velocity Under Loaded Conditions

Based on the initial record of angular displacement of the forearm as a function of time, the maximal linear velocity (meters per second) was determined in the following manner: The slope of the tangent line of the angular displacement/time curve was calculated at an angle of 85° relative to the horizontal. This one angle was used because, in all records analyzed, the slope of the tangent line became the steepest and was a straight line in the angular range of 80° to 90°. Hence, angular velocity was maximal and constant within this range. The steepness of the slope of the tangent line was obtained by placing

[1]Our appreciation is extended to Dr. Thomas Szirtes of the Biomedical Engineering Unit for his contribution to the construction of the ergometer.

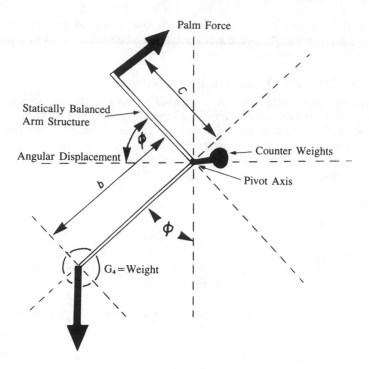

Figure 1. Arrangement of the ergometer (see text for explanation).

a front surface reflecting mirror across the curve at 85° and turned until the reflection of the curve formed a straight line with the graph. A line was drawn along the edge of the mirror with a sharp pencil, and the angle of this line was measured with a protractor. This angle was subtracted from 90 and the tangent of the subsequent angle was calculated. The maximal angular velocity was therefore represented by the equation:

$$\Phi = 100 \tan (90 - \Phi)$$

(I), where

$$\Phi = \text{the maximal angular velocity (degrees per second);}$$
$$100 = \text{(the paper speed of the visicorder (mm per second); and}$$
$$(90 - \Phi) = \text{the slope of the tangent line (degrees)}$$

The maximal angular velocity (degrees per second) was converted to a radian measure by multiplying by 0.175. This value was multiplied by the radius of rotation of the subject's forearm (elbow joint to the center of the palm in centimeters) and converted to its meter equivalent to obtain the maximal linear velocity in meters per second.

Maximal Palm Force Under Loaded Conditions

The maximal palm force (kg) was calculated for an angular displacement of the forearm at an angle of 85° relative to the horizontal, that is, at the same angle at which the maximal angular velocity was obtained. Since all angular velocities were maximal and constant over the range of 80° to 90°, acceleration was zero and the maximal dynamic palm force was represented by the following equation:

$$F_p = \frac{G_4 \times b \times \text{Sine } \Phi}{C}$$

(II), where

F_p = the maximal palm force (kg), acting perpendicular to the resistance arm of the ergometer;

G_4 = the weight (kg) selected for the ergometer that supplied the load on the forearm;

b = the distance between the pivot axis of the ergometer and the center of gravity of the weight on the resistance arm. By geometry and construction of the ergometer, $b = 69.4 - 0.232 \times r$, where r = the reading on a numerical counter which determined b distance;

Φ = the angular displacement of the human forearm with respect to the horizontal (85°); and

C = the distance (cm) between the pivot axis of the elbow joint and the center of the palm of the hand.

One weight (G_4) was previously selected for the twin pair and was either 3.2, 6.5, or 9 kilograms throughout the series of different loads. All twin pairs received identical loads on the forearm flexor muscles. Prior selection of a weight was necessary because if the weight was too heavy for a given set of twins only a small number of different loads could be obtained.

Maximal Linear Velocity (V_{max}) Under Unloaded Conditions

An electrogoniometer was attached with adhesive tape to the lateral side of the forearm and connected to the Honeywell Visicorder Oscillograph. In the experimental starting position of 60° the subject made a maximal voluntary flexion of the forearm throughout the entire range of 50° without any load. The angular displacement of the forearm as a function of time was recorded and the maximal linear velocity under unloaded conditions (V_{max} was obtained by interpolation. Two measurements of V_{max} were made and the highest obtained maximal linear velocity (m per second) was used as the experimental point to V_{max} on the force-velocity curve.

Maximal Isometric Force (P_0)

The lever arm of the ergometer was rigidly fixed at an angle of 80° relative to the horizontal. Isometric force has been shown to be greatest at this angle, and previous investigators have used this angle when P_0 values were determined for force-velocity curves of human forearm flexor muscles (Ikai & Fukunaga, 1970). A force dynamometer (kg) was inserted in the lever arm of the ergometer and adjusted to the forearm length of the subject. Two maximal voluntary contractions of the forearm flexors were obtained with a 3-minute rest period between the trials. The highest value attained was used as the experimental P_0 point on the force-velocity curve.

Maximal Muscular Power

The maximal force and velocity values, in addition to the maximal velocity of movement under unloaded conditions (V_{max}) and the maximal isometric force (P_0), provided the experimental points on which the force-velocity curve was drawn. For each twin, the force-velocity curve was drawn as the curve that best fit the experimentally obtained values. In only four twins, of a total number of 34 force-velocity curves drawn, was the obtained maximal isometric force (P_0) less than what was predicted from the shape of the curve. In these cases the force-velocity curve was extrapolated to a point that predicted the P_0 value from the shape of the curve.

Once the force-velocity curves were drawn, the maximal muscular power was calculated for each twin by obtaining the product of two variables, the maximal force (kg) and its corresponding maximal linear velocity (m per second) for successive time intervals along the force-velocity curve. The product was multiplied by 60 and the maximal muscular power (kg = m per minute) was plotted and drawn on the same graph as the force-velocity curve.

Experimental Procedure

Each subject made four practice trials with the highest load at different speeds in order to familiarize himself with the movement of the lever arm. The subject was instructed to make a single maximal voluntary flexion of the forearm, rest 3 minutes, and then attempt another maximal voluntary flexion at a slightly heavier load. This procedure was continued until the subject was unable to move the load over the entire range of 50°.

When the trial under loaded conditions ended, an electrogoniometer was taped to the lateral side of the forearm and was placed in the starting position of 60°. When the subject was ready, the visicorder was turned on and the maximal velocity of movement (V_{max}) under unloaded conditions was recorded. Another trial was taken after a 3-minute rest period.

The frame structure of the ergometer was then bolted in such a way that the lever arm remained fixed at an angle of 80°. A force dynamometer was inserted in the lever arm and adjusted to the length of the subject's forearm. When ready, the subject made a maximal voluntary contraction against the fixed dynamometer. After 3 minutes rest, another maximal effort was obtained. This procedure was followed for each twin pair.

Results

The anthropometric data and the values obtained for maximal muscular power (MMP), maximal isometric force (P_0), and maximal velocity of movement (V_{max}) of MZ and DZ twins are presented in Table 1. Individual force-velocity curves for co-twin pairs along with the corresponding muscular power curves are shown in Figures 2 and 3 for MZ and DZ twins, respectively. Inspection of the force-velocity-power curves indicates that the intrapair differences are greater between DZ pairs than MZ pairs.

Within-pair variance estimates for MMP, P_0, and V_{max} were generated separately for MZ and DZ twins by a one-way analysis of variance. The within-pair variance F-ratio was significant ($p < 0.01$) for MMP and P_0, but not for V_{max}. Thus, heritability estimates were computed only for MMP and P_0, using the Holzinger index. The respective estimates were 97% and 83%.

Discussion

The individual force-velocity curves obtained from the twins have followed the well known exponential form, while the maximal muscular power curves, derived from these curves, have added support to the notion that the maximal power of the forearm flexor muscles results when force and velocity values are approximately 35% of the maximum values (Komi, 1973). The accuracy of the obtained results for maximal muscular power (MMP) depends upon the precision with which the force-velocity curves were drawn to best fit the experimentally obtained values. Only two sets of twins revealed a lower maximal isometric force (P_0) than that predicted by extrapolation of the force-velocity curve to zero velocity. Fatigue factors, brought about by the great number of different loads that were used, may have contributed to the lower observed values for P_0. In two cases only six loads were used because the previously selected weight was too heavy and only a limited range of loads was available. In these cases the accuracy of the force-velocity-power curves may be questioned. However, all other force-velocity curves were based on a substantial number of different loads which made the drawing of the best fit curve precise and acceptable.

Inspection of Figures 2 and 3 indicates that the intrapair differences for MMP are greater between DZ pairs than MZ pairs, and that the greater intrapair differences for DZ twins are a direct result of greater individual differences in the relative position of the force-velocity curve.

Table 1. Means, standard deviations, and range of measurements on monozygous (MZ) and dizygous (DZ) twins

Variables	MZ twins			DZ twins		
	M	SD	Range	M	SD	Range
Age (yrs)	13.4	±1.6	4	12.5	±1.9	6
Height (cm)	156.3	±9.6	28.2	150.3	±9.8	32
Weight (kg)	48.0	±9.9	34	48.1	±10.8	22.7
V_{max}(m/sec)	3.12	±0.56	2.33	2.81	±0.48	1.85
P_0 (kg)	16.2	±4.6	16.5	12.51	±3.37	12.0
MMP(kg·m/min)	323.1	±183.5	637.4	211.1	±120.6	408.2

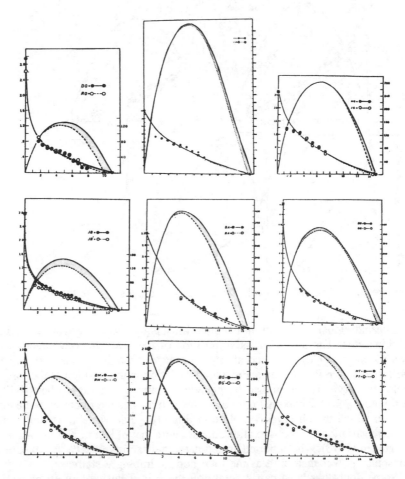

Figure 2. Individual force-velocity and power curves for MZ twins.

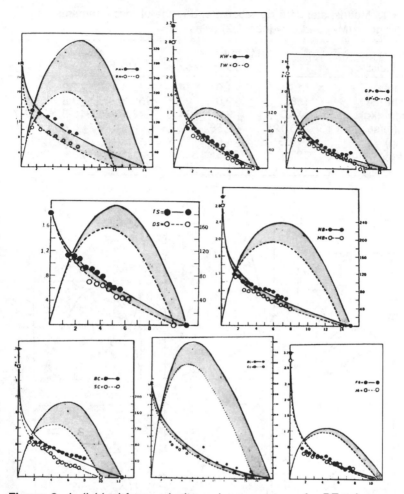

Figure 3. Individual force-velocity and power curves for DZ twins.

Interindividual variability in maximal muscular power is governed by individual differences in the relative position of the force-velocity curve (Ikai & Fukunaga, 1970). The heritability estimate for maximal muscular power will therefore reflect the extent to which variation in the position of the force-velocity curves is determined by genetic variation. The higher heritability estimate (97%) for maximal muscular power indicates the rather extreme extent to which the observed variation in this polygenic variable is subject to genetic variation. This observation concurs with that of Komi, Klissouras, and Karvinen (1973), who reported an estimated heritability of 99% for maximal muscular power measured as the total body climbing a staircase.

Some doubt has been cast recently on the use of heritability estimates in studying the relative share of genetic and nongenetic influences in interindividual differences of polygenic attributes (Kang, Christian, & Norton, 1978; Bouchard & Lortie, 1984). In this regard it should be pointed out that the validity of the heritability estimate depends upon the acceptability of the under-lying assumptions. Four basic assumptions were made in the derivation of the estimate. It was assumed, first, that environmental influences are comparable for both MZ and DZ twins; second, that hereditary variance shows no dominance or interaction effects; third, that no correlation exists between parents due to assortative mating; and fourth, that genetic and nongenetic in-fluences are not correlated. The tenability of accepting these assumptions has been previously discussed (Klissouras, 1971). In order to strengthen the ac-ceptance of these assumptions, further emphasis in the present study was placed on obtaining comparable MZ and DZ groups with regard to socioeconomic backgrounds, leisure time activities, age, and physical endeavors.

However, there is some contradictory evidence with regard to the genotype-environment interaction (Weber, Kartodihardjo, & Klissouras, 1976; Prud'homme, Bouchard, Landry, Fontaine, D'Amours, & Leblanc, 1983; Boulay, Lortie, Simoneau, & Bouchard, 1986). If the response of polygenic attributes to environmental influences is genotype dependent, the heritability index should be modified to include an additional term signifying the interac-tion between heredity and environment. However, until such evidence is available for maximal muscular power, heritability estimates derived from the present data may be reasonable.

References

Bouchard, C., & Lortie, G. (1984). Heredity and endurance performance. *Sports Medicine*, **1**, 38-64.

Boulay, M.R., Lortie, G., Simoneau, J.-A., & Bouchard, C. (1986). Sensitivity of maximal aerobic power and capacity to anaerobic training is partly genotype depen-dent. In R.M. Malina & C. Bouchard (Eds.), *Sport and human genetics* (pp. 173-182). Champaign, IL: Human Kinetics.

Ikai, M., & Fukunaga, T. (1970). A study of training effect on strength per unit cross-sectional area of muscle by means of ultra-sonic measurement. *Internationale Zeitschrift für Angewandte Physiologie*, **28**, 173-180.

Kang, K.W., Christian, J.C., & Norton, J.A., Jr. (1978). Heritability estimates from twin studies. I. Formulae of heritability estimates. *Acta Geneticae Medicae et Giemellogiae*, **27**, 39-44.

Klissouras, V. (1971). Heritability of adaptive variation. *Journal of Applied Physiology*, **31**, 338-344.

Komi, P.V. (1973). Measurement of the force-velocity relationship in human mus-cle under concentric and eccentric contractions. In S. Cerquiglini, A. Venerando, & J. Wartenweiler (Eds.), *Biomechanics III* (pp. 224-229). Baltimore: University Park Press.

Komi, P.V., Klissouras, V., & Karvinen E. (1973). Genetic variation in neuro-muscular performance. *Internationale Zeitschrift für Angewandte Physiologie*, **31**, 289-304.

Prud'homme, D., Bouchard, C., Landry, F., Fontaine, E., D'Amours, Y., & Leblanc, C. (1983). Sensitivity of maximal aerobic power is genotype dependent. *Medicine and Science in Sports and Exercise*, **15**, 133 (abstract).

Weber, G., Kartodihardjo, W., & Klissouras, V. (1976). Growth and physical training with reference to heredity. *Journal of Applied Physiology*, **40**, 211-215.

10
Anaerobic Alactacid Work Capacity in Adopted and Biological Siblings

Jean-Aimé Simoneau, Gilles Lortie,
Claude Leblanc, and Claude Bouchard
LAVAL UNIVERSITY
QUEBEC, CANADA

Muscular alactacid anaerobic power, that is, the power output associated with the ATP-PC high energy system, has been investigated in several laboratories since the advent of the test devised by Margaria, Aghemo, and Rovelli (1966). The data available in the literature have clearly demonstrated that maximal alactacid anaerobic power in humans shows considerable interindividual variation. Age, sex, body weight, relative muscle mass, and physical activity habits are among the potential factors contributing to maximal alactic anaerobic working power of humans (di Prampero, 1981; Margaria et al., 1966, Whiters, McFarland, Cousins, & Gore, 1979). Moreover, among all causal sources contributing to variation, heredity is generally viewed as exerting a significant influence. Indeed, Komi and Karlsson (1979) showed significant intrapair differences in the Margaria test between male MZ and DZ twins.

It is now generally recognized, however, that the study of anaerobic energy systems requires not only distinction between the alactacid and lactacid components but also between the capacity and power of each component (Bouchard, Thibault, & Jobin, 1981). It is well known that intramuscular phosphagens (ATP-PC) are responsible for the production of short-term supramaximal physical work lasting 8 to 10 seconds or less (di Prampero, 1981). During an all-out bicycle ergometer test, the external mechanical power increases during the first 2 to 3 seconds of exercise and reaches a maximum level which is maintained up to the 7th second (di Prampero & Mognoni, 1981). The test described by Margaria et al. (1966) appears to adequately reflect the alactacid anaerobic power. For measuring the anaerobic alactacid capacity, the choice of a 10-second test, as in the present study, was supported by the fact that the lactacid mechanism does not contribute significantly to the energy production during the first 10 seconds of an exhaustive exercise (Åstrand, 1981). However, there is little information on the anaerobic alactacid work capacity (AAC). Therefore, the purpose of this study was to quantify the degree of sibship similarity in AAC in order to gain some insight about the effect of the genotype on this measurement.

Methods

Subjects

A total of 328 individuals from families of French descent participated in this study. They belonged to adopted sibships ($n = 19$ sibships, 15.6 ± 3.0 [mean $\pm SD$] years of age), biological sibships of regular brothers and sisters ($n = 55$, 18.7 ± 5.6), DZ twin sibships ($n = 31$, 15.3 ± 2.6), and MZ twins ($n = 49$, 18.2 ± 5.6). The subjects ranged between 9 and 33 years of age, the majority being between 12 and 25 years.

Anaerobic Alactacid Capacity (AAC)

AAC was determined as the total work output that could be performed in 10 seconds, that is, using the Quebec AAC test. Following a 5-minute warm-up on a cycle ergometer, a 10-minute mandatory rest period was imposed before the first test. After the first trial and a 10-minute resting period, a second trial was imposed. All tests were performed on a modified Monark ergocycle equipped with a photoelectric cell registering each third of a flywheel revolution and a potentiometer connected to the tension adjustment mechanism of the ergocycle determining the work load. The highest work output in joules (J) during the better of two trials was retained for analysis. More detailed procedures for the test and its reliability have been previously described (Simoneau, Lortie, Boulay, & Bouchard, 1983).

Fat-Free Weight

The percent body fat was determined from body density obtained through underwater weighing. Procedures for determining body density were as described by Behnke and Wilmore (1974). Residual volume was assessed according to the method employed by Wilmore, Vodak, Parr, Girandola, and Billing (1980). Percent body fat was estimated from body density according to Siri (1956). Fat-free weight was computed from percent of body fat and actual body weight.

Statistical Procedures

To control for the effects of age and sex on AAC measurements, multiple regressions were applied (age + sex + age X sex + age^2). Residual scores of AAC measurements were computed and submitted to analysis of variance and correlation analysis. The analysis of variance and the F ratio computed from the between-sibships over the within-sibship variance were obtained following the procedures outlined in Haggard (1958). Equality of mean and of total variance according to twin type was tested following the procedures described by Christian (1979).

Results

Means, standard deviations, and indices of skewness and of kurtosis for all subjects are presented in Table 1. Average age, weight, and percent body fat were 17.2 years (SD = 4.8), 53.0 kg (SD = 13.6), and 15.7% (SD = 7.5), respectively. The average values for AAC/kg of body weight and AAC/kg of FFW were 86.9 J/kg and 101.6 J/kg FFW, respectively. The size of the standard deviation relative to mean value indicates considerable variation for both AAC measurements. Multiple regression performed on AAC measurements revealed that age and sex accounted for 58% and 47% of the variance in AAC/kg and AAC/kg FFW, respectively.

Results of the analysis of variance and intraclass correlations for sibs performed with residuals of age and sex are presented in Table 2. While the F ratios of between-sibship over within-sibship means of squares were signifi-

Table 1. Descriptive statistics and anaerobic alactacid working capacity of adopted and biological sibships[a]

Variable	N	M	SD	Skewness	Kurtosis
Age (years)	328	17.2	4.8	0.8	0.7
Body weight (kg)	328	53.0	13.6	0.1	−0.2
Percent fat	314	15.7	7.5	0.4	−0.3
AAC/kg (J/kg)	328	86.9	21.3	0.4	−0.3
AAC/kg FFW (J/FFW)	314	101.6	20.4	0.3	−0.2

[a]Age and sex accounted for 58% of the variance in AAC/kg and 47% in AAC/kg FFW.

Table 2. Intraclass coefficients and ratios of between-sibships over within-sibship means of squares for alactacid anaerobic work capacity measurements[a]

Sibships	N Sibships	AAC/KG		AAC/kg FFW	
		r	F ratio	r	F ratio
Adoptive sibships	19	−0.01	0.98	0.06	1.14
Biological sibships	55	0.46	2.33[b]	0.38	2.45*
DZ sibships	31	0.58	3.89[b]	0.44	2.61*
MZ sibships	49	0.80	9.15[b]	0.77	7.75*

[a]On residuals of age and sex.
*$p < 0.01$.
[b]$p < .01$.

cant for the biological, the DZ, and the MZ sibships, significance was not reached in the adopted sibships. The greatest sibship resemblance was observed for MZ twins, whereas biological and DZ sibship resemblances were quite similar. Intraclass coefficients did not differ from zero in adopted sibs, whereas they ranged from 0.38 to 0.46 in biological sibships, from 0.44 to 0.58 in DZ sibships, and from 0.77 to 0.80 in MZ sibships.

Tests for the homogeneity of means and variances between DZ and MZ twins are presented in Table 3. The mean and total variance for all AAC measurements were identical in each twin type. Moreover, the within-component estimate of genetic variance was significant for all measurements. Broad heritability estimates obtained from twice the biological sib correlations or from twice the difference between MZ and DZ correlations are also presented in Table 3. Broad heritability estimates for AAC measurements ranged from 0.76 to 0.92 with the sib method, and from 0.44 to 0.66 with the twin procedure.

Discussion

Results of this study indicate that about 50% of the variance in AAC per unit body weight within a population sample 9 to 33 years of age is associated with age and sex of subjects. These observations agree with those of other studies. Thus, Margaria et al. (1966) observed that alactacid anaerobic work power increases with age to reach a maximum between 20 and 30 years. Moreover,

Table 3. Comparison of means and variances by twin type. Test for significance of genetic variance and broad heritability estimates from twin and sibling data[a]

Variable	AAC/kg	AAC/kg FFW
DZ vs MZ mean difference[b] t ratio	0.99	1.71
DZ vs MZ equality of variance[b] F ratio	1.07	1.32
Genetic variance[b] Gwt	2.00*	1.91*
Broad heritability estimates sibs[c] twins[c]	0.92 0.44	0.76 0.66

[a]On residuals of age and sex.
[b]As described by Christian (1979).
[c]From Falconer (1960).
*$p < .05$.

differences between males and females have been reported by Cumming (1972) and Simoneau et al. (1983). In the latter investigation, the performance of females was approximately 80% that of males when AAC was expressed in Joules per kg of body weight.

Alactacid anaerobic work capacity has not been previously studied in adoptive or biological sibships. In the present study, adopted children did not exhibit similarity in anaerobic alactacid capacity. Intraclass coefficients in households of adopted individuals living together ranged from −0.01 to 0.06. For biological sibships, significant resemblances were found for both AAC measurements. Intraclass coefficients between pairs of brothers and sisters living together varied from 0.38 to 0.46. These results suggest that sharing about one-half of the genome and living together translate into an enhanced resemblance over that found in cohabitating sibs who have no genes in common by immediate descent.

Correlations between biological sibs and DZ twins were quite similar for AAC measurements, suggesting that increased environmental similarity (as is the case when contrasting DZ twins and regular brothers and sisters) does not translate into increased phenotypic resemblance for AAC per kg and per kg FFW.

Using intrapair differences observed in MZ and DZ twins, Komi, Klissouras, and Karvinen (1973) found an intrapair variance in DZ twins that was about five times higher than in MZ twins for a measurement of maximal anaerobic power (Margaria test) in males. Intraclass coefficients computed from the data available in Komi et al.'s original paper reached 0.85 and 0.98 for the DZ and MZ twin sibships, respectively. However, these results were based on an indicator of anaerobic work power with no control over body weight. Again, intraclass coefficients computed from the data in the original paper for the mechanical power output, expressed in kpm per kg of body weight per sec, reached 0.69 and 0.80 for DZ and MZ twin sibships, respectively. These latter results are similar to those obtained for AAC/kg in the present study. The intraclass coefficients in this study were 0.58 and 0.80 for the DZ and MZ sibships, respectively (Table 2).

Values previously reported concerning the total genetic effect in maximal anaerobic power are widely divergent, ranging from almost zero to almost 100% of the variance (Komi et al., 1973; Komi & Karlsson, 1979). These results must be interpreted with caution because sample sizes, differential effects of age and sex according to twin types, and differences in means and variances between the twin populations are factors that may not have been considered in the analysis (Bouchard & Malina, 1983). In the present study, however, such precautions were taken in order to eliminate as many biases as possible.

With the data base of the present study the total genetic effect in anaerobic alactacid capacity can be estimated with various procedures, such as the within-component estimate of heritability (Christian, 1979), the coefficient of twice the difference between the DZ and MZ intraclass coefficients, or twice the correlations between pairs of biological sibs (Falconer, 1960). The genetic variance calculated as described by Christian revealed that AAC/kg and AAC/kg FFW are both characterized by a significant genetic variance component. Depending upon the computational procedures used, the total genetic effects ranged from 0.44 to 0.76, with one exception for AAC/kg from the sib method (broad heritability estimate = .92). It is quite clear that the heritability estimates derived from the sib data in the present study tend to be higher than those obtained from the twin data. One should not be surprised with this finding, as heritability estimates based on sib data alone tend to be inflated. Common lifestyle and physical activity pattern effects are not controlled when using twice the intraclass coefficient from biological sibships in comparison with the twin procedure.

In conclusion, resemblance between members of a sibship is highest when they share genes by descent. AAC/kg and AAC/kg FFW are both characterized by a significant genetic variance component. Broad heritability estimates derived from biological sibs or from twin data suggest that a major portion of the variance in anaerobic alactacid capacity is accounted for by a genetic effect.

References

Åstrand, P.O. (1981). Aerobic and anaerobic energy sources in exercise. *Medicine and Sport*, **13**, 22-38.

Behnke, A.R., & Wilmore, J.H. (1974). *Evaluation and regulation of body build and composition*. Englewood Cliffs, NJ: Prentice Hall.

Bouchard, C., & Malina, R.M. (1983). Genetics of physiological fitness and motor performance. *Exercise and Sport Sciences Reviews*, **11**, 306-339.

Bouchard, C., Thibault, M.C., & Jobin, J. (1981). Advances in selected areas of human work physiology. *Yearbook of Physical Anthropology*, **24**, 1-36.

Christian, J.C. (1979). Testing twin means and estimating genetic variance. Basic methodology for the analysis of quantitative twin data. *Acta Geneticae Medicae et Gemellologiae*, **28**, 35-40.

Cumming, G.R. (1972). Correlation of athletic performance and aerobic power in 12- to 17-year-old children with bone age, calf muscle, total body potassium, heart volume and two indices of anaerobic power. In O. Bar-Or (Ed.), *Proceedings of the fourth international symposium of pediatric work physiology* (pp. 109-134). Israel: Wingate Institute.

di Prampero, P.E. (1981). Energetics of muscular exercise. *Review of Physiology, Biochemistry and Pharmacology*, **89**, 143-222.

di Prampero, P.E., & Mognoni, P. (1981). Maximal anaerobic power in man. *Medicine and Sport*, **13**, 38-44.

Falconer, D.S. (1960). *Introduction to quantitative genetics*. New York: Ronald Press.

Haggard, E.A. (1958). *Intra-class correlation and the analysis of variance*. New York: Dryden Press.

Komi, P.V., & Karlsson, J. (1979). Physical performance, skeletal muscle enzyme activities and fiber types in monozygous and dizygous twins of both sexes. *Acta Physiologica Scandinavica*, (Suppl. 462), 1-28.

Komi, P.V., Klissouras, V., & Karvinen, E. (1973). Genetic variation in neuromuscular performance. *Internationale Zeitschrift fur Angewandte Physiologie*, **31**, 289-304.

Margaria, R., Aghemo, P., & Rovelli, E. (1966). Measurement of muscular power (anaerobic) in man. *Journal of Applied Physiology*, **21**, 1662-1664.

Simoneau, J.A., Lortie, G., Boulay, M.R., & C. Bouchard. (1983). Test of anaerobic alactacid and lactacid capacities: Description and reliability. *Canadian Journal of Applied Sport Sciences*, **8**, 266-270.

Siri, W.E. (1956). The gross composition of the body. In J.H. Lawrence & C.A. Tobias (Eds.), *Advance in biological and medical physics* (pp. 239-280). New York: Academic Press.

Whiters, R.T., McFarland, K., Cousins, L., & Gore, S. (1979). The measurement of maximum anaerobic alactacid power in males and females. *Ergonomics*, **22**, 1021-1028.

Wilmore, J.H., Vodak, P.A., Parr, R.B., Girandola, R.N., & Billing, J.E. (1980). Further simplification of a method for determination of residual lung volume. *Medicine and Science in Sports and Exercise*, **12**, 216-218.

11

Sensitivity of Maximal Aerobic Power and Capacity to Anaerobic Training is Partly Genotype Dependent

Marcel R. Boulay, Gilles Lortie, Jean-Aimé Simoneau, and Claude Bouchard
LAVAL UNIVERSITY
QUEBEC, CANADA

The study of biological changes induced by exercise training in humans is an important topic in the sport sciences. Maximal aerobic power (MAP) has often been considered in the study of responses to moderately intense training regimes. Mean increments for MAP generally reported in such training experiments range from about 10% to 40% of the pretraining value (Åstrand & Rodahl, 1970; Bouchard, Carrier, Boulay, Thibault-Poirier, & Dulac, 1975), while individual responses vary considerably, for example between 5% and 88% in a recent study in our laboratory (Lortie, Simoneau, Hamel, Boulay, Landry, & Bouchard, 1984). These wide variations strongly suggest that all individuals do not respond in the same manner to a standardized aerobic training program. The effects of high intensity-short duration (interval, anaerobic) training, on the other hand, have also been investigated. This mode of training seems to increase MAP to a lesser degree, that is, from about 5% to 24% in several studies (Cohen & Gisolfi, 1982; Cunningham & Faulkner, 1969; Fox, 1975; Fox, Bartels, Billings, O'Brien, Bason, & Mathew, 1975; Fox, Bartels, Klinzinc, & Ragg, 1977; Knuttgen, Nordesjö, Ollander, & Saltin, 1973; Lesmes, Fox, Stevens, & Otto, 1978).

Thanks are expressed to L. Pérusse, H. Bessette, C. Leblanc, G. Fournier, and P. Hamel for their assistance in the course of this study. This study was supported by NSERC (G-0850 and A-8150) and FCAC-Québec (EQ-1330).

Individual differences in the response of MAP to exercise training raise the question of the role of heredity in adaptability (Bouchard, 1983; Bouchard & Lortie, 1984). Prud'homme, Bouchard, Leblanc, Landry, and Fontaine (1984b) reported that improvements in MAP (ml O_2/kg•min^{-1}) following a 20-week aerobic training program were characterized by a significant ($r_i = 0.74$) level of resemblance within MZ twin pairs, thus suggesting that the sensitivity to training may be genotype dependent. This study considers the genotype-training interaction in responses of MAP, maximal aerobic capacity (MAC), and ventilatory threshold to a 15-week anaerobic training program.

Methods

Subjects

Fourteen pairs of MZ twins of both sexes (7 female and 7 male pairs) gave their informed written consent to participate in this study. Zygosity was established by questionnaire, with several red blood-cell antigen and enzyme genetic markers, and with the A, B, and C loci of the HLA system. All subjects were known to be sedentary or only slightly active (male twins) prior to the experiment. Age, body weight, and percent body fat were 21.2 ± 3.3 years (mean ± SD), 58.2 ± 10.0 kg, and 18.6 ± 6.6%, respectively. All procedures were approved by the Medical Ethics Committee of Laval University.

Maximal Aerobic Power Test (MAP)

A MAP test was conducted on an electromagnetically braked ergocycle. The initial power output was 50 watts, and increments of 20 watts for women and 25 for men were made every 3 minutes until exhaustion. Gas exchanges were continuously monitored with an automated open circuit system (MMC, Beckman). The heart rate (HR) was obtained with an ECG recording in the CM5 lead position.

Ventilatory Threshold (VT-2)

The second ventilatory threshold (McLellan, 1983) was visually determined as the second nonlinear increase in $\dot{V}e$ and/or $\dot{V}e/\dot{V}o_2$ relative to $\dot{V}O_2$. VT-2 was assessed by two independent evaluators as described by Prud'homme, Bouchard, Leblanc, Landry, Lortie, and Boulay (1984a).

Maximal Aerobic Capacity Test (MAC)

MAC was measured as the total work output performed during a 90-minute maximal ergocycle test (Boulay, Hamel, Simoneau, Lortie, Prud'homme, & Bouchard, 1984), that is, the Quebec aerobic capacity test. The test was performed on a modified Monark ergocycle (Simoneau, Lortie, Boulay, & Bouchard, 1983). The starting work load was calculated from the results of the MAP test in order to elicit a HR approximately 10 beats lower than that observed at VT-2. The work load was continuously adjusted throughout the test in order to maintain the highest intensity that could be sustained by the subject.

Measurement of Body Fatness

Body density was determined through underwater weighing. Subjects were strapped with a weighted diving belt to ensure submersion, and measurement was performed after a moderate inspiration. The mean of six valid measurements was used in calculating body density. Procedures for determining body density were those described by Behnke and Wilmore (1974). Percent fat was estimated from body density with the Siri (1956) equation. The method employed by Wilmore, Vodak, Parr, Girandola, and Billing (1980) was used to determine residual lung volume.

Training Protocol

Subjects were put on a 15-week cycle ergometer training program that used both continuous and interval work patterns. The 25 continuous, 19 short, and 16 long interval sessions were unevenly distributed during the program, such that half of the continuous sessions were completed by the 5th week of training. This was done to minimize the occurrence of overuse problems in these untrained subjects, which could have occurred if high-intensity interval work had been imposed too soon. The characteristics of the program are described in Table 1.

Statistical Analysis

Differences between means were tested for significance with the paired t-test. Intraclass coefficients were computed following the procedures outlined in Haggard (1958).

Results

There were no changes in body weight or body density in the total sample of twins. However, there were small but significant decreases in body weight and in percentage body fat, and a slight increase in fat-free mass in male twins. There were no significant changes in females.

The effects of the training program on MAP, MAC, and VT-2 are shown in Table 2. Training increased MAP/kg by 22%, MAC/kg by 17%, and VT-2/kg by 34% in the total sample of twins. All changes were significant ($p < 0.001$). Individual changes, however, ranged from about zero to 65% for MAP/kg, from about zero to 55% for MAC/kg, and from 6% to 91% for VT-2/kg.

Table 3 shows mean individual changes for subgroups of male and female subjects. The results for MAC, MAC/kg, and VT-2/kg indicated similar increments in both sexes. Increments in MAP and MAP/kg, although similar in $1 \cdot min^{-1}$ or $ml/kg \cdot min^{-1}$, represented different percentages of pretraining values. However, the relative changes did not differ significantly between the sexes.

Figure 1A illustrates the resemblance in within-twin pair response of MAC/kg to the training program. The intraclass coefficient (r_i) computed from the magnitude of changes in MAC/kg expressed as a percentage of the first test reached 0.69. The Pearson correlation coefficient (r) computed between the

Table 1. Characteristics of the training program

Mode of training		Weeks of training														
		1	2	3	4	5	6	7	8	9	10	11	12	13	14	15
Continuous	Number of sessions	3	3	3	2	2	2	1	2	2	1	1	1	1	1	1
	Intensity[a] (%)	65-70	65-75	70	70	70/75	70	70	70/75	70/80	70	70	70	70	70	70
	Duration (min)	20-25	30	30	30	35	30/35	35	30/35	30	30	30	30	30	30	30
Short interval	Number of sessions	0	0	1	2	1	1	1	2	1	2	1	2	2	1	1
	Intensity[b] (%)			60	60	60	65	65	70	70	70	70	75	75/80	75	80
	Duration (sec)			15	15	15/30	15/20	15/20	15/30	15	15/30	15	15/20	15	15	15
	Repetitions			15	15/18	8/5	10/5	8/7	10/8	10	10/8	12	10	10	10	10
Long interval	Number of sessions	0	0	0	0	1	2	2	1	1	2	2	1	1	2	1
	Intensity[c] (%)					70	75	80	80	80	85	85	90	90	85/90	90
	Duration (sec)					60	60	70	80	85	80	80	90	90	80	90
	Repetitions					5	5	4	5	4	5	5	3	4	5	4

[a]% of maximal heart rate reserve (MHRR = maximal HR – resting HR; Karvonen et al., 1957).
[b]% of individual maximal work output in 10 sec. (Simoneau et al., 1983).
[c]% of individual maximal work output in 90 sec. (Simoneau et al., 1983).

Table 2. Effects of a 15-week anaerobic training program on maximal aerobic power (MAP), maximal aerobic capacity (MAC), and ventilatory threshold (VT-2) (N = 28)

	Variable	First test	Second test	Individual changes as percentage of first test
MAP	$(1 \cdot min^{-1})$	2.45 ± 0.71[a]	2.93 ± 0.72*	22.1 ± 14.6
	$(m10_2/kg \cdot min^{-1})$	41.5 ± 6.8	50.2 ± 6.8*	22.4 ± 13.4
	(kJ)	677 ± 195	786 ± 218*	17.0 ± 12.3
MAC				
	(kJ/Kg)	11.46 ± 2.24	13.22 ± 2.23*	16.5 ± 13.6
VT-2	$(m10_2/kg \cdot min^{-1})$	34.1 ± 5.9	43.3 ± 7.2*	34.1 ± 18.5

[a]Mean ± SD.
*$p < 0.001$.

Table 3. Mean changes in males and females induced in maximal aerobic power (MAP), maximal aerobic capacity (MAC), and ventilatory threshold (VT-2) by a 15-week anaerobic training program

	Variable	Individual changes Males	Individual changes Females	Individual changes as percentage of pretraining values Males	Individual changes as percentage of pretraining values Females
MAP	$(1 \cdot min^{-1})$	0.50	0.47	17.6	26.7
	$(m10_2/kg \cdot min^{-1})$	8.3	9.3	18.5	26.2
	(kJ)	137	80	17.7	16.3
MAC					
	(kJ/Kg)	2.2	1.4	18.7	14.4
VT-2	$(m10_2/kg \cdot min^{-1})$	11.6	10.8	31.8	36.5

improvements in MAC/kg and its pretraining level was -0.47, that is, 22% common variance; it is shown in Figure 1B.

Figures 2 and 3 illustrate the results for MAP/kg and VT-2/kg, respectively. For these variables, intraclass coefficients were 0.44 (MAP/kg) and -0.06 (VT-2/kg), while Pearson correlation coefficients with pretraining levels were -0.56 (31%) and -0.47 (22%).

Discussion

The magnitude of changes observed in MAP/kg in the present study is in the upper range of values reported in the literature for anaerobic training programs.

For instance, the increments reported in MAP/kg for the short-interval training program in male subjects ranged from 5% (Fox, 1975; Fox et al., 1977)
to 15% (Knuttgen et al., 1973), while Lesmes et al. (1978) reported an increase of 12% in female subjects. For long-interval training in male subjects,
reported improvements ranged from 5% (Fox, Bartels, Billings, Mathew,
Bason, & Webb, 1973) to 24% (Knuttgen et al., 1973); Cohen and Gisolfi
(1982) reported an 11% increment in females. Furthermore, with a training
program using both short and long intervals, Fox et al. (1975) reported a 14%
gain of MAP in males. In the present study MAP/kg increased by 22% when
the total samples were considered. Differences in intensity and repetition of
interval training bouts, as well as the total duration of the training program,
may explain the rather large increase in MAP/kg in the present study.

Increments in MAC/kg and VT-2/kg were, respectively, smaller and greater
than those observed in MAP/kg, an observation that is consistent with the

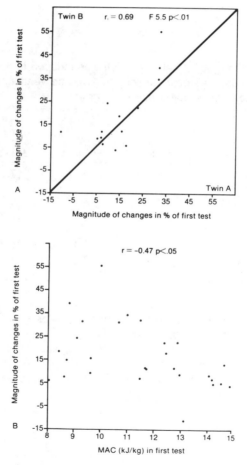

Figure 1. Intraclass resemblance (intraclass coefficient) in the magnitude of
training changes in MAC/kg expressed as a percentage of the pretraining
value.

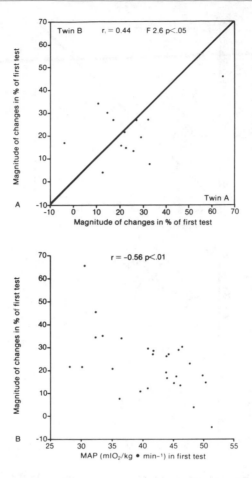

Figure 2. Intrapair resemblance (intraclass coefficient) in the magnitude of training changes in MAP/kg expressed as a percentage of the pretraining value.

specificity concept of training. Continuous sessions were not very intense and were purposely grouped at the beginning of the program, while the interval sessions required energy expenditure greater than VT-2/kg energy output level. Gains in MAP/kg and MAC/kg are contrary to those reported by Lortie et al. (1984), who observed greater MAC and lesser MAP improvements following a 20-week endurance training program, as one could have predicted.

The adaptive responses resulting from the training program in this study were highly variable, as indicated by interindividual differences for changes in MAC/kg, MAP/kg, and VT-2/kg. Such variations have been reported previously with aerobic training (Bouchard, 1983; Bouchard et al., 1975; Lortie et al., 1984; Prud'homme et al., 1984b). Moreover, the intraclass coefficients computed with the training responses indicated that these variations in MAP/kg and MAC/kg were not distributed randomly in the twin sample. The sensitivity

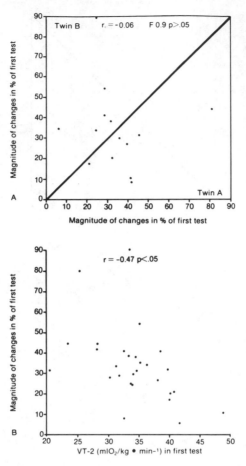

Figure 3. Intrapair resemblance (intraclass coefficient) in the magnitude of training changes in VT-2/kg expressed as a percentage of the pretraining value.

to training for these two variables was significantly similar in members of the same twin pair, but it was not for VT-2/kg.

Bouchard and Malina (1983) have proposed that sex and age of subjects, previous training experiences, and current phenotype level as well as specific genetic variations could be some causal factors associated with differential responses to training. The subjects of the present study did not vary greatly in age (range from 16 to 26), so age could possibly be eliminated. Sex seems to exert a moderate effect on the trainability of MAP and VT-2, while MAC may be more susceptible, as shown earlier by Lortie et al. (1984) for endurance training.

The results of the present study do not support the hypothesis of significant sex differences in the response to anaerobic training, even though females appeared to have higher gains in MAP and MAP/kg expressed as a percentage of the pretraining level. The relatively low scores obtained by female twins in the first test (MAP/kg = 37) compared to male twins (MAP/kg = 46), in association with relatively similar gains in both sexes, can explain the higher percentage increases observed in females. Estimates of trainability in males may also have been altered slightly by their tendency toward an active mode of living. However, past experience in anaerobic training was not a contributing factor in this study, and no subjects had trained with an ergocycle. The initial phenotype level and specific genetic variation thus appear to be variables most likely to contribute to the observed differences in response to the training program.

The intraclass coefficient represents the extent of the within-pair resemblance in contrast to the total variation observed. The F ratios indicated that 69% and 44% of MAC/kg and MAP/kg changes, respectively, were associated with genotype differences. These results suggest that responses to anaerobic training in MAP/kg seemed less associated with the genotype than when endurance training was used (Prud'homme et al., 1984b). On the other hand, within this genotype-training interaction component a fraction is associated with the initial level of the variable. It is well known that the relationship between initial $\dot{V}O_2$ max and the gain obtained with aerobic training is significant. Bouchard et al. (1975) reviewed 50 studies dealing with aerobic training available in the literature and found that about 25% of the improvement observed in MAP/kg was accounted for by the initial $\dot{V}O_2$ max. The present study yielded coefficients of the same order not only for MAP/kg, but also for MAC/kg and VT-2/kg.

In summary, the present study confirms previous findings from our laboratory (Prud'homme et al., 1984b) that the response of aerobic performance to exercise training is partially genotype dependent. The data clearly indicate that the sensitivity of maximal aerobic power and maximal aerobic capacity to high intensity intermittent exercise training is related to the genotype of the individuals. However, some of the variation in response to anaerobic training is associated with the pretraining level of aerobic performance, which is also partially inherited.

References

Åstrand, P.O., & Rodahl, K. (1970). *Textbook of work physiology*. New York: McGraw-Hill.

Behnke, A.R., & Wilmore, J.H. (1974). *Evaluation and regulation of body build and composition*. Englewood Cliffs, NJ: Prentice-Hall.

Bouchard, C. (1983). Human adaptability may have a genetic basis. In F. Landry (Ed)., *Health risk estimation, risk reduction and health promotion. Proceedings of the 10th annual meeting of the Society of Prospective Medicine* (pp. 463-476). Ottawa: Canadian Public Health Association.

Bouchard, C., Carrier, R., Boulay, M.R., Thibault-Poirier, M.C., & Dulac, S. (1975). *Le développement du système de tranport de l'oxygène chez les jeunes adultes*. Québec: Editions de Pélican.

Bouchard, C., & Lortie, G. (1984). Heredity and endurance performance. *Sports Medicine*, **1**, 38-64.

Bouchard, C., & Malina, R.M. (1983). Genetics for the sport scientist: Selected methodological considerations. *Exercise and Sport Sciences Reviews*, **11**, 275-305.

Boulay, M.R., Hamel, P., Simoneau, J.-A., Lortie, G., Prud'homme, D., & Bouchard, C. (1984). A test of aerobic capacity: Description and reliability. *Canadian Journal of Applied Sport Sciences*, **9**, 122-126.

Cohen, J.S., & Gisolfi, C.V. (1982). Effects of interval training on work-heat tolerance of young women. *Medicine and Science in Sports and Exercise*, **14**, 46-52.

Cunningham, D.A., & Faulkner, J.A. (1969). The effect of training on anaerobic and anaerobic metabolism during a short exhaustive run. *Medicine and Science in Sports*, **1**, 65-69.

Fox, E.L. (1975). Differences in metabolic alterations with sprint versus endurance interval training programs. In H. Howald & J.R. Poortmans (Eds.), *Metabolic adaptation to prolonged physical exercise* (pp. 119-126). Basel, Switzerland: Birkhäuser Verlag.

Fox, E.L., Bartels, R.L., Billings, C.E., Mathew, D.K., Bason, R., & Webb, W.M. (1973). Intensity and distance of interval training programs and changes in aerobic power. *Medicine and Science in Sports*, **5**, 18-22.

Fox, E.L., Bartels, R.L., Billings, C.E., O'Brien, R., Bason, R., & Mathew, D.K. (1975). Frequency and duration of interval training programs and changes in aerobic power. *Journal of Applied Physiology*, **38**, 481-484.

Fox, E.L., Bartels, R.L., Klinzinc, J., & Ragg, K. (1977). Metabolic responses to interval training programs of high and low power output. *Medicine and Science in Sports*, **9**, 191-196.

Haggard, E.A. (1958). *Intra-class correlation and the analysis of variance*. New York: Dryden Press.

Karvonen, M.J., Kentola, E., & Mustola, O. (1957). The effects of training on heart rate: A longitudinal study. *Annales Medicinae Experimentalis et Biologae Fenniae*, **33**, 307-315.

Knuttgen, H.G., Nordesjö, L.O., Ollander, B., & Saltin, B. (1973). Physical conditioning through interval training with young male adults. *Medicine and Science in Sports*, **5**, 220-226.

Lesmes, G.R., Fox., E.L., Stevens, C., & Otto, R. (1978). Metabolic responses of females to high intensity interval training of different frequencies. *Medicine and Science in Sports*, **10**, 229-232.

Lortie, G., Simoneau, J.A., Hamel, P., Boulay, M.R., Landry, F., & Bouchard, C. (1984). Responses of maximal aerobic power and capacity to aerobic training. *International Journal of Sports Medicine*, **5**, 232-236.

McLellan, T.M. (1983). Ventilatory and plasma lactate response with different exercise protocols: A comparison of methods. *Canadian Journal of Applied Sports Sciences*, **8**, 214. (abstract)

Prud'homme, D., Bouchard, C., Leblanc, C., Landry, F., Lortie, G., & Boulay, M.R. (1984a). Reliability of assessments of ventilatory thresholds. *Journal of Sports Sciences*, **2**, 13-24.

Prud'homme, D., Bouchard, C., Leblanc, C., Landry, F., & Fontaine, E. (1984b). Sensitivity of maximal aerobic power to training is genotype dependent. *Medicine and Science in Sports and Exercise*, **16**, 489-493.

Simoneau, J.A., Lortie, G., Boulay, M.R., & Bouchard, C. (1983). Tests of anaerobic alactacid and lactacid capacities: Description and reliability. *Canadian Journal of Applied Sport Sciences*, **8**, 266-270.

Siri, W.E. (1956). The gross composition of the body. *Advances in Biological and Medical Physics*, **4**, 239-280.

Wilmore, J.H., Vodak, P.A., Parr., R.B., Girandola, R.N., & Billing, J.E. (1980). Further simplification of a method for determination of residual lung volume. *Medicine and Science in Sports and Exercise*, **12**, 216-218.

Concluding Remarks

This volume constitutes a concerted effort by scientists, most of them working independently, to summarize the available literature and introduce new data concerning the genetics of physical fitness particularly as it relates to sport performance. Actually, vast portions of the field are not even dealt with in this volume because so few authors and laboratories are actively involved in the study of biological inheritance as it pertains to sport.

It is never a simple task to undertake the study of sport, from any point of view, even when one tries to limit the scope to sport performance. Indeed, sport performance cannot be defined as a unique and homogeneous phenotype. In other words, sport performance encompasses a variety of complex phenotypes, and within a particular sport, as in team sports, specialized roles or given positions are associated with different biological, mechanical, perceptual, and skill requirements resulting in increased heterogeneity of sport performance phenotypes. As a consequence, research about the genetics of sport performance is and will undoubtedly remain a difficult undertaking.

Performance in any sport is not a discrete trait, i.e., either present or absent; rather, it is a phenotype measured or qualified on a continuous scale with a large number of achievement classes (seconds, kilograms, meters, goals scored, points, etc.). Performance is thus a quantitative trait influenced by a variety of agents such as training, nutrition, weather conditions, age of the individual, motivation, and even chance variation. Performance is also determined by innate biological, mechanical, and psychological factors. Thus, sport performances qualify readily for what is known in genetic epidemiology as multifactorial traits.

Most studies in the area of genetics and performance have dealt with known determinants of performance such as body size, body fat, muscular strength, muscular endurance, speed of reaction and of movement, skeletal muscle fiber type distribution and enzyme activity, cardiac dimensions, pulmonary volumes, and maximal aerobic power. Little has been reported on the genetics of sport performance itself considered as the phenotype of interest, except for a few early studies concerned with the question of whether outstanding sport performance occurred in family lines (Bouchard & Malina, 1984).[1] In general, the studies described in this volume are not unlike those early efforts. By means of various research strategies using twins, nuclear families, or relative by adoption, the authors investigated whether there was evidence for a significant genetic effect in some of the biological determinants of sport performance.

What we have learned from the Eugene symposium on Human Genetics and Sport is exciting, even if limited, when considered in the broader context of sport performance.

1. It has been reasserted that the genotype is clearly a major force operating on physiological fitness and health-related conditions.
2. It has been demonstrated that under normal conditions, children's physical growth is largely determined by genetic factors in terms of both the level of growth achieved and the speed of that growth.
3. It has been shown that motor performance tasks are generally influenced by a moderate but significant genetic effect, with perhaps a higher heritability in males than in females.
4. It has been established that maximal aerobic power and capacity, anaerobic alactic capacity, and skeletal muscle histochemical and biochemical characteristics are characterized by only a moderate heritability component.
5. It has been confirmed that one of the most striking genetic effects is not the average genetic effect measured in the population, but rather the role of the genotype in determining the rate and the amplitude of the response to a chronic stimulation such as exercise training.
6. It has been confirmed that there are high responders and low responders for maximal aerobic power and capacity in response to either endurance training or high-intensity intermittent training, and it has been proposed that the respective frequency of the high or low responder phenotypes is approximately 5% of the young adult population.

It is also important to note that no studies of causal factors in these genetic effects were reported in the symposium. In other words, no data on transcription, or enzyme and other protein polymorphisms in relevant tissues, or on the mitochondrial genome in relation to sport performance were presented. Even though these concepts were sometimes involved in the discussion, the ubiquitous influences of consanguinity, heterosis, racial differences, and racial admixture have not been systematically considered in the context of the athlete and sport performance. Hopefully, the next Olympic Scientific Congress will be able to pull together enough resources to permit an in-depth discussion of the old and new concepts and to draw an even more comprehensive picture of the involvement of the genome in sport performance.

Claude Bouchard
Robert M. Malina
Editors

[1]Bouchard, C., & Malina, R.M. (1984). Genetics and Olympic athletes: A discussion of methods and issues. In J.E.L. Carter (Ed.), *Kinanthropometry of Olympic Athletes* (pp. 28-38). Basel: Karger.